Fermentation Biotechnology

ACS SYMPOSIUM SERIES **862**

Fermentation Biotechnology

Badal C. Saha, Editor

Agricultural Research Service
U.S. Department of Agriculture

Sponsored by the
ACS Division of Biochemical Technology

American Chemical Society, Washington, DC

Library of Congress Cataloging-in-Publication Data

Fermentation biotechnology / Badal C. Saha, editor.

   p. cm.—(ACS symposium series ; 862)

   "Sponsored by the ACS Division of Biochemical Technology."

   Includes bibliographical references and index.

   ISBN 978-0-8412-3845-9

   1. Fermentation—Congresses.

   I. Saha, Badal C., 1949- II. American Chemical Society. (224th : 2002 ; Boston, Mass.) III. Series.

TP156.F4F47   2003
660.28449—dc21   2003052400

The paper used in this publication meets the minimum requirements of American National Standard for Information Sciences—Permanence of Paper for Printed Library Materials, ANSI Z39.48–1984.

PRINTED IN THE UNITED STATES OF AMERICA

Foreword

The ACS Symposium Series was first published in 1974 to provide a mechanism for publishing symposia quickly in book form. The purpose of the series is to publish timely, comprehensive books developed from ACS sponsored symposia based on current scientific research. Occasionally, books are developed from symposia sponsored by other organizations when the topic is of keen interest to the chemistry audience.

Before agreeing to publish a book, the proposed table of contents is reviewed for appropriate and comprehensive coverage and for interest to the audience. Some papers may be excluded to better focus the book; others may be added to provide comprehensiveness. When appropriate, overview or introductory chapters are added. Drafts of chapters are peer-reviewed prior to final acceptance or rejection, and manuscripts are prepared in camera-ready format.

As a rule, only original research papers and original review papers are included in the volumes. Verbatim reproductions of previously published papers are not accepted.

ACS Books Department

Contents

Overview

Production of Specialty Chemicals

Production of Pharmaceuticals

Environmental Remediation

Metabolic Engineering

Process Validation

Indexes

Preface

Tremendous advances have been made in fermentation biotechnology for the production of a wide variety of commodity chemicals and pharmaceuticals. It is timely to provide a book that can assist practicing scientists, engineers, and graduate students with effective tools for tackling the future challenges in fermentation biotechnology.

This book was developed from a symposium titled *Advances in Fermentation Process Development,* presented at the 224th National Meeting of the American Chemical Society (ACS) in Boston, Massachusetts, August 18–22, 2002 and sponsored by the ACS Division of Biochemical Technology. It presents a compilation of seven symposium manuscripts and eight solicited manuscripts representing recent advances in fermentation biotechnology research. The chapters in the book have been organized in five sections: Production of Specialty Chemicals, Production of Pharmaceuticals, Environmental Bioremediation, Metabolic Engineering, and Process Validation. An overview chapter on commodity chemicals production by fermentation has been included.

I am fortunate to have contributions from world-class researchers in the field of fermentation biotechnology. I am taking this opportunity to express my sincere appreciation to the contributing authors, the reviewers who provided excellent comments to the editor, the ACS Division of Biochemical Technology, and the ACS Books Department for making possible the publication of this book.

I hope that this book will actively serve as a valuable multi-disciplinary (biochemistry, microbiology, molecular biology, and biochemical engineering) contribution to the continually expanding field of fermentation biotechnology.

Badal C. Saha
Fermentation Biotechnology Research Unit
National Center for Agricultural Utilization Research
Agricultural Research Service
U.S. Department of Agriculture
1815 North University Street
Peoria, IL 61604
(309) 681–6276 (telephone)
 (309) 681–6427 (fax)
sahabc@ncaur.usda.gov (email)

Fermentation Biotechnology

Overview

Chapter 1

Commodity Chemicals Production by Fermentation: An Overview

Badal C. Saha

Fermentation Biotechnology Research Unit, National Center
for Agricultural Utilization Research, Agricultural Research Service,
U.S. Department of Agriculture, Peoria, IL 61604

Various commodity chemicals such as alcohols, polyols, organic acids, amino acids, polysaccharides, biodegradable plastic components, and industrial enzymes can be produced by fermentation. This overview focuses on recent research progress in the production of a few chemicals: ethanol, 1,3-propanediol, lactic acid, polyhydroxyalkanoates, exopolysaccharides and vanillin. The problems and prospects of cost-effective commodity chemical production by fermentation and future directions of research are presented.

During the last two decades, tremendous improvements have been made in fermentation technology for the production of commodity chemicals and high value pharmaceuticals. In addition to classical mutation, selection, media design, and process optimization, metabolic engineering plays a significant role in the improvement of microbial strains and fermentation processes. Classical mutation includes random screening and rationalized selection. Rationalized selection can be based on developing auxotropic strains, deregulated mutants, mutants resistant to feedback inhibition and mutants resistant to repression (*1*). In addition to the classical approach to media design and statistical experimental design, evolutionary computational methods and artificial neural networks have been employed for media design and process optimization (*1*). Important regulatory mechanisms involved in the biosynthesis of fermentation products by a microorganism include substrate induction, feedback regulation, and nutritional regulation by sources of carbon, nitrogen, and phosphorus (*2*). Various metabolic engineering approaches have been taken to produce or improve the production of a metabolite by fermentation (*3*). These are: (i) heterologous protein production, (ii) extension of substrate range, (iii) pathway leading to new products, (iv) pathways for degradation of xenobiotics, (v) engineering of cellular physiology for process improvement, (vi) elimination or reduction of by-product formation, and (vii) improvement of yield or productivity.

As the demand for bio-based products is increasing, attempts have been made to replace more and more traditional chemical processes with faster, cheaper, and better enzymatic or fermentation methods. Significant progress has been made for fermentative production of numerous compounds such as ethanol, organic acids, calcium magnesium acetate (CMA), butanol, amino acids, exopolysaccharides, surfactants, biodegradable polymers, antibiotics, vitamins, carotenoids, industrial enzymes, biopesticides, and biopharmaceuticals. Fermentation biotechnology contributes a lot to the pollution control and waste management. This chapter gives an overview of the recent research and developments in fermentation biotechnology for production of certain common commodity chemicals by fermentation.

Ethanol

Ethanol has widespread application as an industrial chemical, gasoline additive, or straight liquid motor fuel. In 2002, over 2 billion gallons of ethanol were produced in the USA. The demand for ethanol is expected to rise very sharply as a safer alternative to methyl tertiary butyl ether (MTBE), the most common additive to gasoline used to provide cleaner combustion. MTBE has been found to contaminate ground water. Also, there is increased interest to replace foreign fossil fuel with a much cleaner domestic alternative fuel derived from renewable resources.

Currently, more than 95% of fuel ethanol is produced in the USA by fermenting glucose derived from corn starch. In the USA, ethanol is made from corn by using both wet milling and dry milling. In corn wet milling, protein, oil, and fiber components are separated before starch is liquefied and saccharified to glucose which is then fermented to ethanol by the conventional yeast *Saccharomyces cerevisiae*. In dry milling, ethanol is made from steam cooked whole ground corn by using simultaneous saccharification and fermentation (SSF) process. Ethanol is generally recovered from fermentation broth by distillation.

Both processes are mature and have become state of the art technology. However, various waste and underutilized lignocellulosic agricultural residues can be sources of low-cost carbohydrate feedstocks for production of fuel ethanol. Lignocellulosic biomass generates a mixture of sugars upon pretreatment itself or in combination with enzymatic hydrolysis. *S. cerevisiae* cannot ferment other sugars such as xylose and arabinose to ethanol. Some yeasts such as *Pachysolen tannophilus*, *Pichia stipitis*, and *Candida shehatae* ferment xylose to ethanol (*4, 5*). These yeasts are slow in xylose fermentation and also have low ethanol tolerance (*6, 7*). It is not cost-effective to convert xylose to xylulose using the enzyme xylose-isomerase which can be fermented by *S. cerevisiae* (*8, 9*). Only a few yeast strains can hardly ferment arabinose to ethanol (*10, 11*). Yeasts are inefficient in the regeneration of the co-factor required for conversion of arabinose to xylulose. Thus, no naturally occurring yeast can ferment all these sugars to ethanol.

Some bacteria such as *Escherichia coli*, *Klebsiella*, *Erwinia*, *Lactobacillus*, *Bacillus*, and *Clostridia* can utilize mixed sugars but produce no or limited quantity of ethanol. These bacteria generally produce mixed acids (acetate, lactate, propionate, succinate, etc.) and solvents (acetone, butanol, 2,3-butanediol, etc.). Several microorganisms have been genetically engineered to produce ethanol from mixed sugar substrates by using two different approaches: (a) divert carbon flow from native fermentation products to ethanol in efficient mixed sugar utilizing microorganisms such as *Escherichia*, *Erwinia*, and *Klebsiella* and (b) introduce the pentose utilizing capability in the efficient ethanol producing organisms such as *Saccharomyces* and *Zymomonas* (*12-15*). Various recombinant strains such as *E. coli* KO11, *E. coli* SL40, *E. coli* FBR3, *Zymomonas* CP4 (pZB5), and *Saccharomyces* 1400 (pLNH32) fermented corn fiber hydrolyzates to ethanol in the range of 21-34 g/L with yields ranging from 0.41-0.50 g of ethanol per gram of sugar consumed (*16*). Martinez et al. (*17*) reported that increasing gene expression through the replacement of promoters and the use of a higher gene dosage (plasmids) substantially eliminated the apparent requirement for large amounts of complex nutrients of ethanologenic recombinant *E. coli* strain. Ethanol tolerant mutants of recombinant *E. coli* have been developed that can produce up to 6% ethanol (*18*). The recombinant *Z. mobilis*, in which four genes from *E. coli*, *xylA* (xylose isomerase), *xylB* (xylulokinase), *tal* (transaldolase), and *tktA* (transketolase) were inserted, grew on xylose as the sole carbon source and produced ethanol at 86% of the theoretical yield (*19*). Deng and Ho (*20*) demonstrated that phosphorylation is a vital step for metabolism of xylose through the pentose phosphate pathway. The gene *XKS1* (encoding xylulokinase) from *S. cerevisiae* and the heterologous genes *XYL1* and *XYL2* from *P. stipitis* were inserted into a hybrid host, obtained by classical breeding of *S. uvarum* and S. *diastaticus*, which resulted in *Saccharomyces* strain pLNH32, capable of growing on xylose alone. Eliasson et al. (*21*) reported that chromosomal integration of a single copy of the *XYL1-XYL2-XYLS1* cassettee in *S. cerevisiae* resulted in strain TMB3001. This strain attained specific uptake rates (g/g.h) of 0.47 and 0.21 for glucose and xylose, respectively, in continuous culture using a minimal medium. Recently, Sedlak and Ho (22) expressed the genes [*arab* (L-ribulokinase), *araA* (L-arabinose isomerase), and *araD* (L-ribulose-5-phosphate 4-epimerase)] from the *araBAD* operon encoding the arabinose metabolizing genes from *E. coli* in *S. cerevisiae,* but the transformed strain was not able to produce any detectable amount of ethanol from arabinose. Zhang et al. (23) constructed one strain of *Z. mobilis* (PZB301) with seven plasmid borne genes encoding xylose- and arabinose metabolizing genes and pentose phosphate pathway (PPP) genes. This recombinant strain was capable of fermenting both xylose and arabinose in a mixture of sugars with 82-84% theoretical yield in 80-100 h at 30 °C. Richard et al. (24) reported that overexpression of all five enzymes (aldose reductase, L-arabinitol 4-dehydrogenase, L-xylulose reductase, xylitol dehydrogenase, and xylulokinase) of the L-arabinose catabolic pathway in *S. cerevisiae* led to growth of *S. cerevisiae* on L-arabinose.

Softwoods such as pine and spruce contain around 43-45% cellulose, 20-23% hemicellulose, and 28% lignin. The hemicellulose contains mainly mannose and 6-7% pentose. *S. cerevisiae* can ferment mannose to ethanol. In Sweden, a fully integrated pilot plant for ethanol production from softwood, comprising both two-stage dilute acid hydrolysis and the enzymatic saccharification process is under construction (25).

Research efforts are directed towards the development of highly efficient and cost-effective cellulase enzymes for use in lignocellulosic biomass saccharification. Also, there is a need for a stable, high ethanol tolerant, and robust recombinant ethanologenic organism capable of utilizing more broad sugar substrates and tolerating common fermentation inhibitors such as furfural, hydroxymethyl furfural, and unknown aromatic acids generated during dilute acid pretreatment.

1,3-Propanediol

1,3-Propanediol (1,3-PD) is a valuable chemical intermediate which is suitable as a monomer for polycondensations to produce polyesters, polyethers, and polyurethanes. It can be produced by fermentation from glycerol by a number of bacterium such as *Klebsiella pneumoniae*, *Citrobacter freundii*, and *Clostridium pastwureunum* (26). It is first dehydrated to 3-hydroxypropionaldehyde which is then reduced to 1,3-PD using $NADH_2$. The $NADH_2$ is generated in the oxidative metabolism of glycerol through glycolysis reactions and results in the formation of by-products such as acetate, lactate, succinate, butyrate, ethanol, butanol, and 2,3-butanediol. Some of the by-products such as ethanol and butanol do not contribute to the $NADH_2$ pool at all. The maximum yield of 1,3-PD (67%, mol/mol) can be obtained with acetic acid as the sole by-product of the oxidative pathway (26). Thus the yield of 1,3-PD depends on the combination and stoichiometry of the reductive and oxidative pathways. Generally, a lower yield is obtained due to conversion of a part of glycerol to cell mass. A variety of culture techniques such as batch culture, fed-batch culture, and continuous cultivation with cell recycle or with immobilized cells have been evaluated for production of 1,3-PD. 1,3-PD concentrations of 70.4 g/L for product tolerant mutant of *C. butyricum* and 70-78g/L for *K. pneummoniae* have been achieved with productivity of 1.5-3.0 g/L.h in fed batch culture with pH control and growth adapted glycerol supply (*26*).

Attempts have been made to produce 1,3-PD from glucose by using two approaches: (i) fermentations of glucose to glycerol and glycerol to 1,3-PD by using a two stage process with two different organisms and (ii) the genes responsible for converting glucose to glycerol and glycerol to 1,3-PD can be combined in one organism (*27, 28*). *S. cerevisiae* produces glycerol from the glycolytic intermediate dihydroxyacetone 3-phosphate using two enzymes - dihydroxyacetone 3-phosphate dehydrogenase and glycerol-3-phosphate phosphatase. Conversion of glycerol to 1,3-PD requires two enzymes - glycerol dehydratase and 1,3-propanediol dehydrogenase. An *E. coli* strain has been constructed containing the genes from *S. cerevisiae* for glycerol production and the genes from *K.pneumoniae* for 1,3-PD production (*29*). The performance of this recombinant strain to convert glucose to 1,3-PD equals or surpasses that of any glycerol to 1,3-PD converting natural organism.

Lactic Acid

Lactic acid (2-hydroxypropionic acid) is used in the food, pharmaceutical, and cosmetic industries. It has the potential of becoming a very large volume, commodity chemical intermediate produced from renewable carbohydrates for use as feedstocks for biodegradable polymers, oxygenated chemicals, environmentally friendly green solvents, plant growth regulators, and specialty chemical intermediates (*30*). A specific stereoisomer of lactic acid (D- or L-form) can be produced by using fermentation technology. Many lactic acid bacteria (LAB) such as *Lactobacillus fermentum*, *Lb. buchneri*, and *Lb. fructovorans* produce a mixture of D- and L-lactic acid (*31*). Some LAB such as *Lb. bulgaricus*, *Lb. coryniformis* subsp. *torquens*, and *Lueconostoc mesenteroides* subsp. *mesenteroides* produce highly pure D-lactic acid and LAB such as *Lb. casei*, *Lb. rhamnosus*, and *Lb. mali* produce mainly L-Lactic acid. The existing commercial production processes use homolactic acid bacteria such as *Lb. delbrueckii*, *Lb. bulgaricus*, and *Lb. leichmonii* (*30*). A wide variety of carbohydrate sources such as molasses, corn syrup, whey, glucose, and sucrose can be used for production of lactic acid. Lactic acid fermentation is product inhibited (*32*). Hujanen et al. (*33*) optimized process variables and concentration of carbon in media for lactic acid production by *Lb. casei* NRRL B-441. The highest lactic acid concentration (118.6 g/L) in batch fermentation was obtained with 160 g glucose per L. Resting *Lb. casei* cells converted 120 g glucose to lactic acid with 100% yield (per L) and a maximum productivity of 3.5 g/L.h. LAB generally require complex rich nutrient sources for growth (*34*). Alternatively, *Rhizopus oryzae* produces optically pure L(+)-lactic acid and can be grown in a defined medium with only mineral salts and carbon sources (*35*). However, low production rate, low yield, and production of significant amounts of other metabolites such as glycerol, ethanol, and fumaric acid are some of the disadvantages of using *R. oryzae* for lactic acid production in comparison with LAB. Recently, Park et al. (*36*) reported efficient production of L(+)-lactic acid using mycelial cotton-like flocs of *R. oryzae* in an air-lift bioreactor. The lactic acid concentration produced by the mycelial flocs in the air-lift bioreactor was 104.6 g/L with a yield of 0.87 g/g substrate using 120 g glucose per L.

Garde et al. (*37*) used enzyme and acid treated hemicellulose hydrolyzate from wet-oxidized wheat straw as substrate for lactic acid production with a yield of 95% and complete substrate utilization by a mixed culture of *Lb. brevis* and *Lb. pentosus* without inhibition. Nakasaki and Adachi (*38*) studied L-lactic acid production from wastewater sludge from a paper manufacturing industry by SSF using a newly isolated *Lb. paracesei* with intermittent addition of cellulase enzyme. The L-lactic acid concentration attained was 16.9 g/L which is 72.2% yield based on the glucose content of the sludge under optimal conditions (at pH 5.0 and 40 °C). Tango and Ghaly (*39*) studied a continuous lactic acid production system using an immobilized packed bed of *Lb. helveticus* and achieved a production rate of 3.9 g/L.h with an initial lactose concentration of 100 g/L and hydraulic retention time of 18 h.

Chang et al. (*40*) used an *E. coli* RR1 *pta* mutant as the host for production of D- or L-lactic acid. A *pta ppc* mutant was able to metabolize glucose exclusively to D-lactate (62.2 g/L in 60 h) under anaerobic conditions and a *pta ldh* mutant harboring the *L-ldh* gene from *Lb. casei* produced L-lactate (45 g/L in 67 h) as the major fermentation product. Dequin and Barre (*41*) reported lactic acid and ethanol production from glucose by a recombinant *S. cerevisiae* expressing the *Lb*. casei L(+)-LDH with 20% of utilized glucose conversion to lactic acid. Porro et al. (*42*) reported the accumulation of lactic acid (20g/L) with productivities up to 11 g/L.h by metabolically engineered *S. cserevisiae* expressing a mammalian *ldH* gene (*ldh-A*). Skory (*43*) showed that at least three different *ldh* enzymes are produced by *R. oryzae*. Two of these enzymes, *ldh*A and *ldh*B, require the cofactor NAD$^+$, while the third enzyme is probably a mitochondrial NAD$^+$- an independent *ldh* used for oxidative utilization of lactate. Recently, Skory (*44*) studied lactic acid production by *S. cerevisiae* expressing the *R. oyzae ldh* gene and reported that the best recombinant strain was able to accumulate up to 38 g lactic acid per L with a yield of 0.44 g/g glucose in 30 h. Dien et al. (*45*) constructed recombinant *E. coli* carrying the *ldh* gene from *Streptococcus bovis* on a low copy number plasmid for production of L-lactate. The recombinant strains (FBR 9 and FBR 11) produced 56-63 g L-lactic acid from 100 g xylose per L at pH 6.7 and 35 °C. The catabolic repression mutants (*ptsG⁻*) of the recombinant *E. coli* strains have the ability to simultaneously ferment glucose and xylose (*46*). The *ptsG⁻* strain FBR19 fermented 100 g sugar (glucose and xylose, 1:1) to 77 g lactic acid per L. Recently, Zhou et al. (*47*) constructed derivatives of *E. coli* W3110 (prototype) as new biocatalysts for production of D-lactic acid. These strains (SZ40, SZ58, and SZ63) require only mineral salts as nutrients and lack all plasmids and antibiotic resistance genes used during construction. D-Lactic acid production by the strains approached the theoretical maximum yield of two molecules per glucose molecule with chemical purity of 98% and optical purity exceeding 99%.

Vaccari et al. (*48*) described a novel system for lactic acid recovery based on the utilization of ion-exchange resins. Lactic acid can be obtained with more than 99% purity by passing the ammonium lactate solution through a cation-exchanger in hydrogen form. Madzingaidzo et al. (*49*) developed a process for sodium lactate purification based on mono-polar and bi-polar electrodialysis at which lactate concentration reached to 150 g/L.

Polyhydroxyalkanoates

Polyhydroxyalkanoates (PHAs) such as poly 3-hydroxybutyric acid (PHB) and related copolymers such as poly 3-hydroxybutyric-co-3-hydroxyvaleric acid (PHB-V) are natural homo- or heteropolyesters (MW 50,000-1,000,000) synthesized by a wide variety of microorganisms such as *Ralstonia eutropha*, *Alcaligenes latus*, *Azotobacter vinelandii*, *Chromobacteruium violaceum*, methylotrophs, and pseudomonads (*50*). These renewable and biodegradable polymers are also sources of chiral synthons since monomers are chirals. PHAs are totally and rapidly degraded to CO_2 and water by microorganisms. They are synthesized when one of the nutritional elements such as N, P, S, O_2, or Mg is limiting in the presence of excess carbon source and accumulated intracellularly to levels as high as 90% of the cell dry weight and act as carbon and energy reserve (*50, 51*). Typically, the strains such as *R. eutropha* and *Bhurkolderia cepacia* are grown aerobically to a high cell density in a medium containing cane sugar and inorganic nutrients (*52*). The cell growth is then shifted to PHB synthesis by limiting nutrients other than carbon source, which is continuously fed at high concentration. After 45-50 h, the dry cell mass contains about 125-150 kg/m³ containing about 65-70% PHB. The cost of PHB production from sucrose has been estimated at $2.65/kg for a 10,000 tons per year plant (*51*). Chen et al. (*53*) developed a simple fermentation strategy for large scale production of poly(3-hydroxy-butyrate-co-3-hydroxyhexanoate) by an *Aeromonas hydrophila* strain in a 20,000 L fermentor using glucose and lauric acid as carbon sources. The bacterium was first grown in a medium containing 50 g glucose per L, and the polyhydroxyalkanoate (PHA) biosynthesis was triggered by the addition of lauric acid (50 g/L) under limited nitrogen or phosphorus condition. After 46 h, the final cell concentration, PHA concentration, PHA content, and PHA productivity were 50 g/L, 25 g/L, 50%, and 0.54 g/l.h, respectively. Lee and Yu (*54*) produced PHAs from municipal sludge in a two-stage bioprocess - anaerobic digestion of sludge by thermophilic bacteria in the first stage and production of PHAs from soluble organic compounds in the supernatant of digested sludge by *A. eutrophus* under aerobic and nitrogen-limited conditions. The PHAs produced accounted for 34% of cell mass, and about 78% of total organic carbon in the supernatant was consumed by the bacterium.

Two approaches can be taken to create recombinant organisms for production of PHAs: (a) the substrate utilization genes can be introduced into the PHA producers and (b) PHA biosynthesis genes can be introduced into a non-PHA producer. Many different recombinant bacteria were developed for enhancing PHA production capacity, for broadening the utilizable substrate ranges, and for producing novel PHAs (*55*). Homologous or heterologous overexpression of the PHA biosynthetic enzymes in various organisms has been attempted. Recombinant *E. coli* strains harboring the *A. eutrophus* PHA biosynthesis genes in a stable high-copy number plasmid have been developed and used for high PHA productivity (*56, 57*). Eschenlauer et al. (*58*) constructed a working model for conversion of glucose to PHBV via acetyl- and propionyl-coenzyme A by expressing the PHA biosynthesis genes from *A. eutrophus* in *E. coli* strain K-12 under novel growth conditions. It is possible to produce PHA from inexpensive carbon sources, such as whey, hemicellulose, and molasses by recombinant *E. coli* (*55*). Liu et al. (*59*) studied the production of PHB from beet molasses by recombinant *E. coli* strain containing the plasmid pTZ18u-PHB carrying *A. eutroplus* PHB biosynthesis genes (*phb*A, *phb*B, and *phb*C) and amphicillin resistance. The final dry cell weight, PHB content, and PHB productivity in a 5 L stirred tank fermentor after 31.5 h fed batch fermentation with constant pH and dissolved O_2 content were 39.5 g/L, 80% (w/w), and 1 g/L.h, respectively. Solaiman et al. (*60*) constructed recombinant *P. putida* and *P. oleovorans* that can utilize triacylglycerols as substrates for growth and PHA synthesis. These organisms produced PHA with a crude yield of 0.9-1.6 g/L with lard or coconut oil as substrate.

Several methods have been developed for the recovery of PHAs (*61*). The most often used method involves extraction of the polymer from the cell biomass with solvents such as chloroform, methylene chloride, propylene carbonate, and dichloroethane. In a non-solvent method, cells were first exposed to a temperature of 80 °C and then treated with a cocktail of various hydrolytic enzymes such as lysozyme, phospholipase, lecithinase, and proteinase. Most of the cellular components were hydrolyzed by these enzymes. The intact polymer was finally recovered as a white powder. High production cost is still a major problem in developing a fermentation process for commercial production of PHA.

Exopolysaccharides

Microbial exopolysaccharides (EPS) can be divided intro two groups: homopolysaccharides such as dextran (*Leu. mesenteroides* subsp. *mesenteroides*), alternan (*Leu. mesenteroides*), pullulan (*Aureobasidium pullulans*), levan (*Z. mobilis*), and β-D-glucans *(Streptococcus sp.)* and heteropolysaccharides such as alginate (opportunistic pathogen *Pseudomonasaeruginosa*), gellan (*Sphingomonas paucimobilis*), and xanthan (*Xanthomonas campestris*). Many species of LAB produce a great variety of EPS with different chemical composition and structure. These EPS contribute to the consistency, texture, and rheology of fermented milk products. The biosynthesis of EPS is complex and requires the concerted action of a number of gene products. Generally, four separate reaction sequences are involved: sugar transport into the cytoplasm, the synthesis of sugar-1-phosphates, activation of and coupling of sugars, and processes involved in the export of the EPS (*62*). EPS production by a LAB is greatly influenced by fermentation conditions such as pH, temperature, oxygen tension, and medium composition. The yields of heteropolysaccharides can vary from 0.150 to 0.600 g/L depending on the strain under optimized culture conditions (*63*). *S. thermophilus* LY03 produced 1.5 g/L heteropolysaccharides when an optimal carbon/nitrogen ratio was used in both milk and MRS media (*64*).

Xanthan gum, which has a wide range of application in several industries, is produced by the bacterium *X. campestris* with a production level as high as 13.5 g/L (*65*). Alginate is a linear copolymer of β-D-mannuronic acid and α-D-guluronic acid linked together by 1,4 linkages. It is widely used as thickeners, stabilizers, gelling agents, and emulsifiers in food, textile, paper making, and pharmaceutical industries. Several bacteria such as *Azotobacter vinelandii* and *P. aeruginosa* produce alginate (*66, 67*). Cheze-Lange et al. (*68*) studied the continuous production of alginate from sucrose by *A. vinelandii* in a membrane reactor. A total of 7.55 g of alginate was recovered from the permeate with a production rate of 0.09g/h, yield of 0.21 g/g sucrose, and specific productivity of 0.022 g/g cell.h.

Vanillin

Vanillin (3-methoxy-4-hydroxybenzaldehyde) is one of the most widely used aroma chemicals in the food industry. It is currently prepared in two ways. Vanillin (US $3200/kg) is extracted from vanilla beans (*Vanilla planifolia*) which contains 2% by weight of it. Pure vanillin (US $13.5/kg) is synthesized from guaiacol. The high price of natural vanillin has stimulated research on developing a bio-based method for production of vanillin.

Ferulic acid [3-(4-hydroxy-3-methoxyphenyl)-propenoic acid] is the major cinnamic acid found in a variety of plant cell walls. Corn fiber contains about 3%

Table 1. Production of some other commodity chemicals by fermentation

Metabolite	Microorganism	Substrate (g/L)	Bioreactor Type	Time (h)	Yield (%)
Sorbitol[a]	*Zymomonas mobilis* (75)	Fructose (325) plus Glucose (325)	Batch	8	91
Erythritol	*Moniliella* sp. (76)	Glucose (300)	Batch	144	37
	Candiada magnoliae (77)	Glucose (400)	Fed batch	-	41
	Torula sp. (78)	Glucose (300)	Fed-batch	88	54
Xylitol	*Candida peltata* (79)	Xylose (50)	Shake-flask	78	56
Glycerol	*Candida glycerinogenes* (80)	Glucose (220)	Batch	72	114 g/L
2,3-Butanediol	*Enterobacter clocae* (81)	Fructose (50)	Shake-flask	39	43
Citric acid	*Candida oleophila* (82)	Glucose	Fed-batch	192	80 g/L
Itaconic acid	*Aspergillus terreus* (83)	Glucose (100)	Shake-flask	225	52
Succinic acid	*Actinobacillus succinogenes* (84)	Glucose			110 g/L
Propionic acid	*Propionibacterium acidipropionici* (85)	Glycerol (20)	Batch	54	12 g/L
Gluconic acid	*Aureobasidium pullulans* (86)	Glucose (350)	Continuous stirred tank	26	74
2-Phenylethanol	*Pichia fermentans* (87)	L-Phenylalanine (1)		16	45
Vitamin B$_{12}$	*Propionibacterium freudenreichii* (88)	Glucose	Anaerobic		206 g/L

[a]The yield is based on fructose present. In addition, the bacterium produces gluconic acid with a yield of 91% based on glucose content.

ferulic acid. Wheat bran is another source of ferulic acid (0.5-1%). Faulds et al. (*69*) developed a laboratory scale procedure to produce free ferulic acid (5.7 g) from wheat bran (1 kg) by using a *Trichoderma* xylanase preparation and *Aspergillus niger* ferulic acid esterase. Using filamentous fungi, a two-stage process for vanillin formation was developed in which a strain of *A. niger* was first used to convert ferulic acid to vanillic acid, which was then reduced to vanillin by a laccase-deficient strain of *Pycnoporus cinnabarinus* (*70*). Shimoni et al. (*71*) isolated a *Bacillus* sp. capable of transforming isoeugenol to vanillin. In the presence of isoeugenol, a growing culture of the bacterium produced 0.61 g/L vanillin (molar yield of 12.4%) and the cell free extract resulted in 0.9 g/L vanillin (molar yield of 14%). Ferulic acid can be converted to isoeugenol by *Nocardia autotrophica* DSM 43100 (*72*). Muheim and Lerch (*73*) found that *Streptomyces setonii* produced vanillin as a metabolic overflow product up to 6.4 g/L with a molar yield of 68% from ferulic acid in shake flask experiments using fed-batch approach.

Lee and Frost (*74*) attempted to generate vanillin from glucose via the shikimate pathway using genetically engineered *E. coli* in a fed-batch fermentation. Strain *E. coli* KL7 with plasmid pKL5.26A or pKL5.97A was used to convert glucose to vanillic acid, which was recovered from the medium and reduced to vanillin by using the enzyme aryl aldehyde dehydrogenase isolated from *Neurospora.crassa*.

Concluding Remarks

Table 1 lists production of some other commodity chemicals by fermentation. Fermentation biotechnology, along with improved downstream processing, has played a great role in the production of bulk chemicals as well as high value pharmaceuticals. It will continue to grow tremendously as more and more pathways have been introduced in microbial hosts. The combination of genetic and process approaches will provide enabling technologies for the production of complex and unexplored chemicals by fermentation in the next decades.

References

1.   Parekh, S.; Vinci, V. A.; Strobel, R. J. *Appl. Microbiol. Biotechnol.* **2000**, *54*, 287-301.
2.   Sanchez, S.; Demain, A. L. *Enzyme Microb Technol.* **2002**, *31*, 895-906.
3.   Nielsen, J. *Appl. Microbiol. Biotechnol.* **2001**, *55*, 263-283.
4.   Schneider, H.; Wang, P. Y.; Chan, Y. K.; Maleszka, R. *Biotechnol. Lett.*

14

1981, 3, 89-92.
5. Bothast, R. J.; Saha, B. C. *Adv. Appl. Microbiol.* **1997**, *44*, 261-286.
6. Du Preez, J. C. *Enzyme Microb. Technol.* **1994**, *16*, 944-956.
7. Hahn-Hagerdal, B.; Jeppsson, H.; Skoog, K.; Prior, B. A. *Enzyme Microb. Technol.* **1994**, *16*, 933-943.
8. Gong, C. S.; Chen, L. F.; Flickinger, M. C.; Chiang, L. C.; Tsao, G. T. *Appl. Environ. Microbiol.* **1981**, *41*, 430-436.
9   Hahn-Hagerdal, B.; Berner, S.; Skoog, K. *Appl. Microbiol. Biotechnol.* **1986**, *24*, 287-293.
10. Saha, B. C.; Bothast, R. J. *Appl. Microbiol. Biotechnol.* **1999**, *52*, 321-326.
11. Dien, B. S.; Kurtzman, C. P.; Saha, B. C.; Bothast, R. J. *Appl. Biochem. Biotechnol.* **1996**, *57/58*, 233-242.
12. Ingram, L. O.; Alterhum, F.; Ohta, K.; Beall, D. S. In: Pierce, G. E. Ed., *Developments in Industrial Microbiology*, **1990**, *31*, 21-30.
13. Ho, N. W. Y.; Chen, Z.; Brainard, A. P. *Appl. Environ. Microbiol.* **1998**, *64*, 1852-1856.
14. Zhang, M.; Eddy, C.; Deanda, K.; Finkelstein, M.; Picataggio. M. *Science* **1995**, 267, 240-243.
15. Hahn-Hagerdal, B.; Wahlborm, C. F.; Gardonyi, M.; van Zyl, W. H.; Cordero Otero, R. R.; Jonsson , L. J. *Adv. Biochem. Eng. Biotechnol.* **2001**, *73*, 53-84.
16. Bothast, R. J.; Nichols, N. N.; Dien, B. S. *Biotechnol. Prog.* **1999**, *15*, 867-875.
17. Martinez, A.; York, S. W.; Yomano, L. P.; Pineda, V. L.; Davis, F. C., Shelton, J. C.; Ingram, L. O. *Biotechnol. Prog.* **1999**, *15*, 891-897.
18. Ingram, L. O.; Aldrich, H. C.; Borges , A. C. C.; Causey, T. B.; Martinez, A.; Morales, F.; Saleh, A.; Underwood, S. A; Yomano, L. P.; York, S. W.; Zaldivar, J.; Zhou, S. *Biotechnol. Prog.* **1999**, *15*, 855-866.
19. Yomano, L. P.; York, S. W.; Ingram, L. O. *J. Ind .Microbiol.* **1998**, *20*, 132-138.
20. Deng , X. X.; Ho, N. W. Y. *Appl. Biochem. Biotechnol.* **1990**, *24/25*, 193-199.
21. Eliasson, A.; Christensson, C.; Wahborn, C. F.; Hahn-Hagerdahl, B. *Appl. Environ. Microbiol.* **2000**, *66*, 3381-3386.
22. Sedlak, M.; Ho, N. W. Y. *Enzyme Microb. Technol.* **2001**, *28*, 16-24.
23. Zhang, M.; Chou, Y.; Picataggio, S.; Finklestein, M. *US Patent* 5,843, 760, **1998**.
24. Richard, P.; Putkonen, M.; Vaananen, R.; Londesborough, J.; Penttila, M. *Biochemistry* **2002**, *41*, 6432-6437.
25. Galbe, M.; Zacchi, G. *Appl. Microbiol. Biotechnol.* **2002**, *59*, 618-628.
26. Zeng, A. P.; Biebl, H. *Adv. Biochem. Eng.* **2002**, *74*, 239-259.
27. Cameron, D. C.; Altaras, N. E.; Hoffman, M. L.; Shaw, A. *J. Biotechnol. Prog.* **1998**, *14*, 116-125.

28. Hartlep, M.; Hussmann, W.; Prayitno, N.; Meynial-Salles, I.; Zeng, A. P. *Appl. Microbiol. Biotechnol.* **2002**, *60*, 60-66.
29. Chotani, G.; Dodge, T.; Hsu, A.; Kumar, M.; LaDuca, R.; Trimbur, D.; Weyler, W.; Sanford, K. *Biochim. Biophys. Acta* **2000**, *1543*, 434-455.
30. Datta, R.; Tsai, S. P. In: Saha, B. C.; Woodward, J. Eds. *Fuels and Chemicals from Biomass*. American Chemical Society, Washington, D. C. **1997**, pp. 224-236.
31. Manome, A.; Okada, S.; Uchimura, T.; Komagata, K. *J. Gen. Appl. Microbiol.* **1998**, *44*, 371-374.
32. Goncalves, L. M. D.; Xavier, A. M.; Reida, J. S.; Carrondo, M. J. T. *Enzyme Microb. Technol.* **1991**, *13*, 314-319.
33. Hujanen, M.; Linko, S.; Linko, Y. Y.; Leisola, M. *Appl. Microbiol. Biotechnol.* **2001**, *56*, 126-130.
34. Hofvendahl, K.; Hahn-Hagerdahl, B. *Enzyme Microb. Technol.* **2000**, *26*, 87-107.
35. Yang, C.; Lu, W.; Tsao, G. T. *Appl. Biochem. Biotechnol.* **1995**, *51/52*, 57-71.
36. Park, E. Y.; Kosakai, Y.; Okabe, M. *Biotechnol. Prog.* **1998**, *14*, 699-704.
37. Garde, A.; Jonsson, G.; Schmidt, A. S.; Ahring, B. K. *Bioresource Technol.* **2002**, *81*, 217-223.
38. Nakasaki, K.; Adachi, T. *Biotechnol. Bioeng.* **2003**, *82*, 263-270.
39. Tango, M. S. A.; Ghaly, A. E. *Appl. Microbiol. Biotechnol.* **2002**, *58*, 712-720.
40. Chang, D.E.; Jung, H. C.; Rhee, J. S.; Pan, J. G. *Appl. Environ. Microbiol.* **1999**, *65*, 1384-1389.
41. Dequin, S.; Barre, P. *Biotechnology* **1994**, *12*, 173-177.
42. Porro, D.; Brambilla, L.; Ranzi, B. M.,; Martegani, E.; Alberghina, L. *Biotechnol. Prog.* **1995**, *11*, 294-298.
43. Skory, C. D. *Appl. Environ. Microbiol.* **2000**, *66*, 2343-2348.
44. Skory, C. D. *J. Ind. Microbiol. Biotechnol.* **2003**, *30*, 22-27.
45. Dien, B. S.; Nichols, N. N.; Bothast, R. J. *J. Ind. Microbiol. Biotechnol.* **2001**, *27*, 259-264.
46. Dien, B. S.; Nichols, N. N.; Bothast, R. J. *J. Ind. Microbiol. Biotechnol.* **2001**, *29*, 221-227.
47. Zhou, S.; Causey, T. B.; Hasona, A.; Shanmugam, K. T.; Ingram, L. O. *Appl. Environ. Microbiol.* **2003**, *69*, 309-407.
48. Vaccari, G.; Gonzalez-Varay R. A.; Campi, A.; Dosi, E.; Brigidi, P.; Matteuzzi, D. *Appl. Microbiol. Biotechnol.* **1993**, *40*, 23-27.
49. Madzingaidzo, L.; Danner, H.; Braun, R. *J. Biotechnol.* **2002**, *96*, 223-239.
50. Madison, L. L.; Huisman, G. W. *Microbiol. Mol. Biol. Rev.* **1999**, *63*, 21-53.
51. Lee, S. Y.; Choi, J. *Polymer Degrad. Stabil.* **1998**, *59*, 387-393.
52. Nonato, R. V.; Mantelatto, P. E.; Rossell, C. E. *Appl. Microbiol. Biotechnol.* **2001**, *57*, 1-5.

53. Chen, G. Q.; Zhang, G.; Park, S. J.; Lee, S. Y. *Appl. Microbiol. Biotechnol.* **2001**, *57*, 50-55.
54. Lee, S.; Yu, J. *Resources, Conservation and Recycling* **1997**, *19*, 151-164.
55. Lee, S. Y.; Choi, J. *Adv. Biochem. Eng.* **2001**, *71*, 183-207.
56. Zhang, H.; Obias, V.; Gonyer, K.; Dennis, D. *Appl. Environ. Microbiol.* **1994**, *60*, 1198-1205.
57. Lee, S. Y.; Yim, K. S.; Chang, H. N.; Chang, Y. K. *J. Biotechnol.* **1994**, 32, 203-211.
58. Eschenlauer, A. C.; Stoup, S. K.; Srienc, F.; Somers, D. A. *Int. J. Biol. Macromolecules* **1996**, *19*, 121-130.
59. Liu, F.; Li, W.; Ridway, D.; Gu, T. *Biotechnol. Lett.* **1998**, *20*, 345-348.
60. Solaiman, D. K. Y.; Ashby, R. D.; Foglia, T. A. *Appl. Microbiol. Biotechnol.* **2001**, *56*, 664-669.
61. Kessler, B.; Weusthuis, R.; Witholt, B.; Eggink, G. *Adv. Biochem. Eng.* **2001**, *71*, 159-1182.
62. Laws, A.; Gu, Y.; Marshall, V. *Biotechnol. Adv.* **2001**, *19*, 597-625.
63. Cerning, J. A.; Marshall, V. M. *Recent Res. Developments in Microbiol.* **1999**, *3*, 195-209.
64. Degeest, B.; De Vuyst, L. *Appl. Environ. Microbiol.* **1999**, *65*, 2863-02870.
65. Rodriguez, H.; Aguilar, L.; Lao, M. *Appl. Microbiol. Biotechnol.* **1997**, *48*, 626-629.
66. Gorin, P. A.; Spencer, J. F. T. *Can. J. Chem.* **1966**, *44*, 993-998.
67. Evans, L. R.; Linker, A. *J. Bacteriol.* **1973**, *16*, 915-24.
68. Cheze-lange, H.; Beunard, D.; Dhulster, P.; Guillochon, D., Caze, A. M.; Morcellet, M.; Saude, N.; Junter, G. A. *Enzyme Microb. Technol.* **2002**, *30*, 656-661.
69. Faulds, C. B.; Bartolome, B.; Williamson, G. *Ind. Crops Prod.* **1997**, *6*, 367-374.
70. Lesage-Meessen, L.; Delattre, M.; Haon, M.; Thibault, J. F.; Colonna Ceccaldi, B.; Brunerie, P.; Asther, M. *J. Biotechnol.* **1996**, *50*, 107-113.
71. Shimoni, E.; Ravid, U.; Shoham, Y. *J. Biotechnol.* **2000**, *78*, 1-9
72. Malarczyk, E.; Koszen-Pileeka, I.; Rogalski, J.; Leonowicz, A. *Acta Biotechnol.* **1994**, 235-241.
73. Muheim, A.; Lerch, K. *Appl. Microbiol. Biotechnol.* **1999**, *51*, 456-461.
74. Lee, K.; Frost, J. W. *J. Am. Chem. Soc.* **1998**, *120*, 10545-10546.
75. Silvereira, M. M.; Wisbeck, E.; Hoch, I.; Jonas, R. *J. Biotechnol.* **1999**, *75*, 99-103.
76. Lin, S. J.; Wen, C. Y.; Liau, J. C.; Chu, W. S. *Proc. Biochem.* **2001**, *36*, 1249-1258.
77. Ryu, Y. W.; Park, C. Y.; Park, J. B.; Seo, J. H. *J. Ind. Microbiol. Biotechnol.* **2000**, *25*, 100-103.

78. Oh, D. K.; Cho, C. H.; Lee, J. K.; Kim, S. Y. *J. Ind. Microbiol. Biotechnol.* **2001,** *26,* 248-252.
79. Saha, B. C.; Bothast, R. J. *J. Ind. Microbiol. Biotechnol.* **1999,** *22,* 633-636.
80. Wang, Z. X.; Zhunge, J.; Fang, H.; Prior, B. A. *Biotechnol. Adv.* **2001,** *19,* 201-223.
81. Saha, B. C.; Bothast, R. J. *Appl. Microbiol. Biotechnol.* **1999,** *52,* 321-326.
82. Anastassiadis, S.; Aivasidis, A.; Wandrey, C. *Appl. Microbiol. Biotechnol.* **2002,** *60,* 81-87.
83. Gyamerah, M. H. *Appl. Microbiol. Biotechnol.* **1995,** *44,* 20-26.
84. Zeikus, J. G.; Jain, M. K.; Elankovan, P. *Appl. Microbiol. Biotechnol.* **1999,** 51, 545-552.
85. Himmi, E. H.; Bories, A.; Boussaid, A.; Hassani, L. *Appl. Microbiol. Biotechnol.* **2000,** *53,* 435-440.
86. Anastassiadis, S.; Aivasidis, A.; Wandrey, C. *Appl. Microbiol. Biotechnol.* **2003,** *61,* 110-117.
87. Huang, C. J. R.; Lee, S. L.; Chou, C. C. *J. Biosci. Bioeng.* **2000,** *90,* 142-147.
88. Bykhovsky, V. Y.; Zaitseev, N. J.; Eliseev, A. A. *Appl. Biochem. Microbiol.* **1998,** *34,* 1-18.

Production of Specialty Chemicals

Chapter 2

Advanced Continuous Fermentation for Anaerobic Microorganism

Ayaaki Ishizaki

New Century Fermentation Research Ltd., Fukuoka 819–0002, Japan

Production of economical and high quality L-lactic acid production for polymer synthesis and cheap ethanol for new formula of gasohol is currently focused with great attention. However, anaerobic microorganisms such as L-lactic acid bacteria and ethanol producing bacteria are unlike common aerobic industrial microorganisms such as yeast for ethanol fermentation and bacteria for glutamic acid fermentation. In many publications, it has been described that chemostat with high cell density accomplished high flux productivity but very high and unstable residual substrate concentration in the spent medium. The author has developed an advanced continuous fermentation system for *Lactococcus,* an L-Lactic acid producer, and *Zymomonas*, an ethanol producer, that can attain very high productivity greater than 3 g/g h of the specific productivity with low residual substrate concentration and fine control. The productivity of this system is more than ten times that of usual aerobic batch and fed batch cultures.

Introduction

For biodegradable polymer, PLA (Poly Lactic Acid) synthesis, it is essential to produce L-Lactic acid of very high stereochemical purity at cheap cost by large scale production process However, the technology for anaerobic fermentation is still in the development stages. Production of lactic acid as chemical industry has not yet been established. Lactic acid microorganisms are anaerobic (although they sometime grow in microaerophilic) growth kinetics of lactic acid bacteria are completely different from those of aerobic microorganism such as the bacteria produce glutamic acid. Development of modern lactic acid fermentation is still in its infancy. Many years of research on our isolate microorganism, *Lactococcus lactis* IO-1, homo-L-lactic acid bacterium (1), its growth kinetics has been characterized. The author has developed the high efficient continuous L-Lactic acid fermentation employing this strain.

Kinetic model of anaerobic fermentation

First, the anaerobic growth characterized by serious end-product inhibition; and sterile cell formation. Therefore, to develop a modern lactic acid fermentation that is for the large scale industrial operation of PLA production likely to petrochemical plastic process, we first must study detail kinetic parameters of the growth of the microorganism and then design an improved L-Lactic acid continuous fermentation system.

End product inhibition

End product inhibition is widely acknowledged in all enzymatic reactions and many metabolic pathways. Serious end product (L-Lactic acid) inhibition was observed in our microorganism as described in the previous report (2). It is known that there are different types of inhibition kinetics, un-competitive, non-competitive, and mixed. However, regardless of these types of inhibition, when Ks value is very small, these three kinetic formulas can be approximated to equation 1,

$$\frac{1}{\mu} = \frac{1}{\mu_{max}} + \frac{L_B}{\mu_{max} K_I} \qquad (1)$$

where μ is specific growth rate, μ_{max} is maximum specific growth rate, L_B is lactate concentration in broth, and K_I is inhibitory coefficient. In the previous work, K_s for strain IO-1 was very small (2). When inhibitor (L-Lactic acid) concentration is within 10 g/l, a linear relationship was obtained between reciprocal of the specific growth rate ($1/\mu$) and the inhibitor concentration (L_B). From this plot, kinetic constants of product inhibition for strain IO-1 were determined as μ_{max}=1.25 1/h, and K_I=5 g/l respectively at 37°C, pH=6.0. However, this linear relationship was lost when the inhibitor concentration became high. A parabola was observed in place of the linear relationship line at L-Lactic acid concentration up to 50 g/l (3). From this result, the growth kinetics of this strain can not be fitted by a simple end product inhibition equation.

Sterile cell formation

To express parabola curve observed, the following relationship was introduced.

$$X = X_0(1 - \alpha)\exp(\mu\,dt) \qquad (2)$$

and μ in above equation is approximated by

$$\mu \approx \frac{\mu_{max}}{1 + \dfrac{L_B}{K_I}} \qquad (3)$$

where X is the viable cell population at time passed dt from time 0, X_0 is the viable cell population at time 0, α is sterile cell formation rate and dt is 1/100 h. An algorithm based on these equations was constructed to calculate the cell population, residual substrate concentration, and product concentration (inhibitor concentration) for every 1/100 h. This computer simulation can draw L-Lactic acid batch fermentation time course (4) with represented α value. A term k, cell decreasing rate, which is the same dimension to μ can be written by using a term α :

$$k = \frac{1}{dt}\ln(1-\alpha). \quad (4)$$

Thus the apparent specific growth rate can be expressed,

$$\mu_a = \mu + k \quad (5)$$

so that equation (1) can be rewritten as

$$\frac{1}{\mu_a} = \frac{1}{\mu_{maX}\left(\dfrac{1}{1+\dfrac{L_B}{K_i}}\right) + k} \quad (6)$$

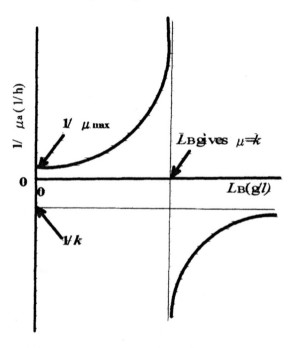

Fig. 1. *Kinetic Model for anaerobic cell growth with end product inhibition.*

In general, anaerobic cell growth is written by

$$X = X_0 \exp(\mu_a t). \quad (7)$$

The maximum cell concentration calculated by computer simulation using different α and the maximum cell concentration determined by turbidity for strain IO-1were compared (3). From this work, α was estimated as 0.0022 (k=0.220 1/h). In the same work, the maximum cell concentration was reached when L_B=23.4 g/l, (μ =0.220 1/h). When μ >0.220 1/h, cell concentration increased and it decreased when μ <0.220 1/h (4). Therefore, to overcome low cell concentration due to sterile cell formation, cell recycling to increase cell concentration in the continuous culture was introduced.

To avoid product inhibition, lower product concentration with high dilution rates was effective (5). To make minimum product inhibition, culture pH is also very important because lactic acid inhibition was only developed by non-dissociated acid (6) so that high culture pHs, its effect becomes reduced. Culture pH of 6.25 gave almost half level inhibition of the culture at pH 6.0

Sterile cell is not a dead cell but that which has weak metabolic activity and has lost its regeneration ability. Although, some enzyme activity still remains, it decreases proportionally (7, 8). This activity decreasing rate can be expressed as

$$E = E_0 \exp\{-0.036(t_0 - t)\} \quad (8)$$

where E_0 is enzyme activity at maximum cell concentration (time= t_0).

Cell immobilization has often been studied as a strategy to enhance the efficiency of many anaerobic fermentation processes. However, this strategy has not succeeded, because sterile cell formation is difficult to eradicate from the immobilization bed. Nonetheless, high activity bioreactor can only be achieved if sterile cells are continuously removed and replaced with fresh (viable) cells.

Substrate feeding strategy

Since aerobic fermentation differs from anaerobic fermentation, dissolved oxygen (DO) level cannot be used as a signal to detect substrate feed in lactic acid fermentation. Moreover, as shown in Fig. 2, during lactic production, there is no established correlation between pH drop and substrate consumption. Therefore, decreasing pH is not a factor for collecting the information for substrate feed. This is because as acid production first decreases, pH goes down according to and then it slowly increases when glucose is almost consumed. However, substrate consumption is still continues after residual glucose concentration exceeds a critical level (Fig. 2). At this point, the cells are almost dead and have lost their viability and no recovery. Therefore, the fermentation is discontinued.

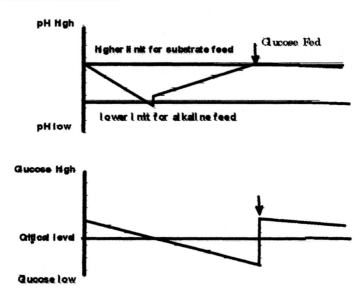

Fig. 2. pH-auxostat can not work for substrate feeding

However, as shown in Fig. 3, if glucose is fed before its residual concentration reaches this critical level, cell activity remains high and the fermentation rate is unaffected. This feeding system is a kind of chemostat although substrate concentration must be maintained above the critical level. Previous works reported very high volumetric productivities using simple chemostat with cell recycling (9). However, in such system, the residual glucose concentration in the spent medium was not controlled and this led to high substrate loss (that is, low product yield) resulting in high production cost; and high residual glucose in the product stream causes poor product quality. Nonetheless, in order to meet the specification of lactic acid for PLA production, the residual glucose in the finished product stream must be lower than 5% of that of L-Lactic acid. To meet this requirement, a strategy for substrate feeding that greatly reduces the residual glucose concentration in the product stream to be the barest minimum must be devised.

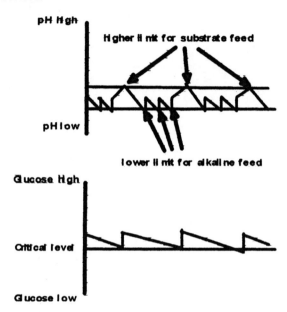

Fig.3. Cell Activity can be maintained when substrate is fed above the critical point..

New concept continuous bioreactor

The author's concept for designing an efficient continuous bioreactor for L-Lactic acid fermentation employing *Lactococcus lactis* IO-1, is as follows:
1. Ease of replacement of sterile cell with fresh cell without using immobilized cells process should be employed;
2 To ensure high cell density, cell recycling process with turbidostat is introduced;
3. Employ high dilution effect to avoid end-product inhibition;
4. Special system for precise control of residual glucose concentration; and
5. Contamination protection tool to facilitate long-time continuous operation.

28

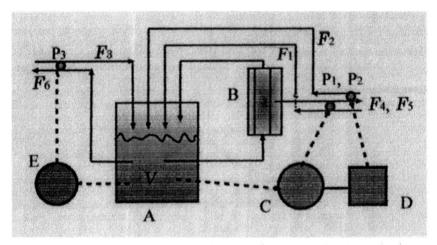

Fig. 4. pH-dependent substrate feed coupled with computer control using cumulative alkali feed and built-in turbidostat.

A: Fermentor, B: Cross Flow Filtration for Cell Separation, C: pH Indicator and Controller, D: Computer, E: Turbidity Controller, F_1: Alkaline Flow-in, F_2: Substrate Flow-in, F_3: Dilute Flow-in F_4 and F_5: Cell Free Broth Exit, F_6: Broth Exit, P1, P2, and P3: Peristaltic Pump

The newly developed high efficient continuous fermentation system has been shown in Fig.4. Since the system operates in continuous mode, to maintain a constant working volume in the fermentor ($dV/dt=0$), the following balance is employed

$$F_1 = F_4, F_2 = F_5, \text{ and } F_3 = F_6 \quad (9)$$

where F_1 is alkaline feed rate (ml/h), F_2; substrate feed rate (ml/h), F_3; dilute feed rate (ml/h), F_4; cell free exit rate balanced for F_1 (ml/h), F_5; cell free exit rate balanced for F_2 (ml/h), F_6; broth bleeding rate (ml/h). This fermentation system consists of a pH-dependent substrate feed system (modified chemostat) coupled with computer-mediated sequence control using cumulative amount of alkali feed to neutralize L-Lactic acid produced and a built-in turbidostat which makes high cell density at the same time works as activity stabilizer (10).

pH-stat works for neutralizing pH caused by L-Lactic acid production. The amount of alkali consumed in pH-stat is equivalent to glucose consumed provided the cell activity in the fermentor is no changed. The amount of glucose required to maintain a constant glucose concentration is therefore given by equation 10.

$$F_2 = \frac{(F_1 \times N \times 90) + \left(\dfrac{F_6 X}{Y_X}\right) + S_R(F_4 + F_5 + F_6)}{S_G} \qquad (10)$$

where N; is normality of alkaline feed solution, Y_X is cell mass yield from glucose (g/g), S_G; glucose concentration of the feed medium (g/l), and S_R; residual glucose concentration in the exit solution (g/l). Thus, glucose demand to refill the consumed glucose (GD g/h) is given by,

$$G_D = S_G \times F_2 + C_G \qquad (11)$$

where C_G is the correction factor for offset of glucose control level monitored by an on-line glucose analyzer.

In the turbidostat, the cell bleeding rate can be expressed by equation 12,

$$\mu_a XV = F_6 X, \text{ thus } \mu_a = D_B = \frac{F_6}{V} \qquad (12)$$

where V is working volume (l) and D_B; s cell bleeding rate (1/h). The total dilution rate of this fermentation system is given by the following equation,

$$D_T = \frac{F_1 + F_2 + F_3}{V} = \frac{F_4 + F_5 + F_6}{V}. \qquad (13)$$

Thus productivity of this system is given by equation 14,

$$P = vX = D_T L_B \qquad (14)$$

where P is the volumetric product productivity (g/l h) and v is the specific product productivity (g/g h).

The above equations suggest that the total dilution rate D_T depends on the volume of the feed solution so that diluted alkali solution and diluted substrate solution can make high dilution effect resulting smaller end product inhibition (11).

Contamination proof system

For polymer synthesis, it is necessary that the L-lactic acid used should be high stereochemical purity. In addition, the levels of residual glucose and contaminated organic acid (that is, fermentation byproduct) in the product stream should be low. Among these technical tasks, stereochemical purity is a major interest because while contaminated organic acids are easily separated by distillation, D-Lactic acid is difficult to remove during the refining process. DL-lactic acid is sometimes produced by foreign microorganisms possess lactate racemase that penetrate into the culture system and converts L-Lactic acid to the DL form. This problem can be prevented if foreign microorganisms kept from the cultivation, refining and recovery steps.

Lactococcus lactis IO-1, homo-L-Lactic acid microorganism is produces nisin Z, a lantibiotic which is bactericidal to gram positive microorganism. Both lactic acid and nisin Z accumulates simultaneously in the broth and therefore potential contaminant microorganisms such as *Bacillus* do not survive in the culture.

In the continuous fermentation system described in the previous section, nisin Z concentration in the culture broth is estimated by;

$$C_B = \frac{v_h X}{D_T} \quad (15)$$

where C_n is nisin Z concentration in broth (IU/) and v_n is specific nisin Z productivity (IU/g h). Our fermentation run gave v_n = 400-500 IU/g h and D_T= 0.3-0.5 l/h, at cell concentration X= about 10 g/l. Thus nisin Z concentration in broth (C_n) is estimated as 8,000= 15,000 IU/l (12). Under such high nisin Z concentration, this fermentation system can last for more than 1,500 h continuous operation without contamination. Such a system can proceed for a long time until mechanical failure occurs. The stereochemical purity (L-Lactic acid of the finishing good was above 99.9%. This special contamination proof system can be applied to the other lactic acid bacterium which produces bacteriocin other than nisin Z.

Summary of continuous fermentation

The results of fermentation run using the system developed are shown in Table 1. As well as *Lactococcus lactis* IO-1, *Zymomonas mobilis* is anaerobic microorganism, and kinetics of cell growth and parameters for fermentation rates are almost the same as strain IO-1. Therefore the system developed for L-Lactic acid production was the same way adapted to ethanol production using *Zymomonas mobilis.* . As shown in Table 1, very efficient continuous process for L-Lactic acid production was confirmed. Volumetric productivity of L-Lactic acid was about 30 g/l h. with lactic acid concentration of 60 g/l and total dilution rate of 0.57 1/h. Specific L-Lactic acid productivity was about 3 g/g h

Table 1. Kinetic parameters of continuous L-Lactic acid fermentation and bacterial ethanol fermentation using new continuous culture system

Operation condition	Microorganism	L-Lactic acid L. lactis IO-1	Ethanol Z. nobilis NRRL B-14023
	pH	6,25	5,5
	Temperature (°C)	37	30
	S_G (g/l)	98	120
	Alkaline conc. (N).	2	
Results	X (g/l)	10.5	8
	L_B (g/l)	59.1	(EtOH) 39.2
	D_T (1/h)	0.57	0.725
	v (g/g h)	3.21	3.55
	P (g/l h)	33.71	28.4
	S_R (g/l)	2.5	0.5
	D_B (1/h)	0.025	0.02
	Product Yield (%)	95	48
	v_n (IU/g h)	400	
	C_n (IU/l)	7,500	
	Time lasted (h)	1,500	800

(13) so far as turbidostat was operating well as an activity stabilizer. The substrate required for refresh of the biomass to replace sterile cell was about 5% of the total substrate feed so that the cost of cell maintenance should be lower than the immobilization cost. Nisin Z concentration was about 7,500 IU/l and no contamination was observed during 1,500 h of continuous operation of the bioreactor.

From the behavior of the alkali feed to *Zymomonas mobilis* in continuous substrate feed, the same reasoning can be adopted to bacterial ethanol production (14, 15). Our previous works show that substrate consumption of this microorganism accompanies pH drop due to proton pump for glucose intake (16). As seen in Table 1, the respective volumetric (30 g/l h) and specific (3 g/g h) productivities of ethanol were almost the same as that obtained for lactic acid. These productivities were higher than those obtained for a whole cell immobilized system of the same microorganism (17).

From this result, this fermentation system can produce about 30 g/l h of the product. For L-Lactic acid production, one short ton of L-Lactic acid per year can be produced by one gallon size fermentor. In the same way, this system can produce 300,000 gallon ethanol per year by 1,000 gallon size fermentor.

Sago Industry

Sago palm is very efficient in photosynthesizing biomass from carbon dioxide, potentially producing 15-20 t of starch per ha per year. This is the highest productivity so far recorded when compared to cereals such as rice, and the other starchy crops (18). This is an extremely high yield. In Sarawak, Malaysia and Riau, Indonesia, sago plantations are under development (19). In the near future, these plantations will be ready for the starch harvesting. Photosynthesis is the best process for the recycling of carbon dioxide from the atmosphere into biomass, and the efficiency of photosynthesis in tropical areas is higher than in moderate and northern areas of the earth.

If biomass can be used in place of petroleum, it will be reduced petroleum consumption. At the same time, carbon dioxide is recycled when the product from biomass decomposes. PLA, biodegradable plastics, must be a good choice for this purpose (20). The new fermentation system can produce 0.72 metric ton of L-lactic acid per day by 1 Kl fermentor. Therefore, about 25,000 ton of L-lactic acid can be produced by one 100 Kl size fermentor in a year. Advanced L-glutamic acid fermentation is possible to accumulate 50-100 g/l of the product in culture liquid for about 2 days batch or fed-batch culture (21). Therefore productivity of glutamic acid fermentation is about 900-1,500 ton per year by one 100 kl size fermentor. Our system is about 20 times of the volumetric productivity of modern L-glutamic acid fermentation. Process yield of L-lactic acid from sago starch is about 100% (105 % from starch to glucose and 95% from glucose to lactic acid so that nearly 100 % for overall process), so that one 100 Kl size fermentor would consume the starch of about 1,000 ha plantation. Fig. 5 shows the model of sago industry that produces PLA from sago plantation. In each 1,000 ha of sago plantation, one fermentation plant with one 100 Kl fermentor of continuous fermentation system is installed. This is similar to an oil palm plantation with an oil mill for crude palm oil extraction.

L-lactic acid fermentation requires nutrition rich organic nitrogen compound such as yeast extract (brewer's waste) and corn steep liquor (waste

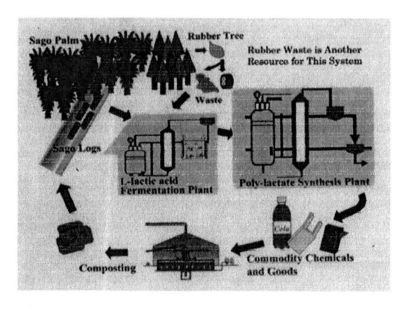

Fig. 5. Sago Industry: PLA production incorporated with Sago plantation and natural rubber plantation for fermentation.

from corn starch process). The author has found that natural rubber waste (waste from latex is excellent nutrition for lactic acid fermentation (22). Sago plantation is sometimes located near by rubber plantation.

Conclusion:

A new concept of continuous fermentation system for anaerobic fermentation has been developed based on the newly established kinetic theory. This system is consists of a modified chemostat coupled to a turbidostat as a cell activity stabilizer and a computer mediated sequence controller using cumulative amount of alkali feed. This system attained very high volumetric productivity as above 30 g of the product per liter hour for L-Lactic acid and ethanol. This modern fermentation process will be introduced to sago palm plantation, newly developing biomass in tropical area to stimulate carbon dioxide recycling and reduce petroleum use.

References:

1. Ishizaki A.; Osajima, K.; Nakamura, K.; Kimura, K.; Hara, T.; Ezaki, T.: *J. Gen. Appl. Microbiol.*, 1990, **36**, 1-6.
2. Ishizaki, A.; Ohta, T.*J. Ferment. Bioeng.*, 1989, **67**, 46-51.
3. Ishizaki, A.; Ohta, T.; Kobayashi, G. *J. Ferment. Bioeng.*, 1989, **68**, 123-130.
4. Ishizaki, A.; Ohta, T.; Kobayashi, G. *J. Biotechnol*, 1992, **24**, 85-107.
5. Ishizaki, A.; Vonktaveesuk, P.: *Biotechnol. Letters*, 1996, **18**. 1113-1118.
6. Yao, P.; Toda, R.: *J. Gen. Appl. Microbiol.*, 1990, **36**, 111-120.
7. Ishizaki, A.; Kobayashi, G. *J. Ferment. Bioeng.*, 1990, **70**, 139-140.
8 Ishizaki, A.; Ueda, T.; Tanaka, K.; Stanbury, P. F.: *Biotechnol. Letters*, 1993, **15**. 489-494.
9. Ohleyer, E.; Wilke, C. R.; Blanch, H. W.: *Appl. Biochem. Biotechnol.*, 1985, **11**, 457-463.
10. Ishizaki, A.: Japanese Patent 2002-087215(not disclosed)
11. Zakpaa, H. D.; Ishizaki, A.: *Biotechnol Techniques*, 1997, **11**,537-541.11. 12.
12. Ishizaki, A.: Japanese Patent 2002-085082 (disclosed)
13. Cirilo, N-H.; Matsunaka, T.; Kobayashi, G.; Sonomoto, K.; Ishizaki, A.; *J. Biosci. Bioeng.*, 2002, **93**, 281-287,.
14. Ishizaki, A.; Tripetchkul, S; Tonokawa, M; Shimizu, K.: *J. Ferment. Bioeng.*, 1994, **77**, 541-547.

15. Zakpaa, H. D.; Ishizaki, A: Biotechnology Techniques, 1997, 11(8), 537-541.;

16. Zakpaa, H. D.; Ishizaki, A.; Shimizu, K. 1997, (B.C. Saha and J. Woodward ed.) ACS Symposium Series, American Chemical Society, Washington DC, 666 Chap. 8, 143-153.

17. Iida, T.; Izumida, H.; Akagi, Y.; Sakamoto, M.: *J. Ferment. Bioeng.*, 1993, 75, 32-35.

18. Ishizaki, A.: The Proceedings for 6th International Sago Symposium. C. Jose et al ed., Riau University, Indonesia, 13.-17. 1996.

19. Jong, F. S.; *Sago Palm* 9(2), 36, 2001

20. Ishizaki, A. 1997, (B.C. Saha and J. Woodward ed.) ACS Symposium Series, American Chemical Society, Washington DC, 666 Chap. 19, 336-344.

21. Kikuchi, M.; Nakao, Y. Production of glutamic acid from sugar. *In* Biotechnology of Amino Acid Production (Aida, K. et al ed.) Elsevier Publishers, Amsterdam, 1986; p103..

22 Tripetchkul, S; Tonokawa, M.; Ishizaki, A.: *J. Ferment. Bioeng.*, 1992, 74, 384-388.

Chapter 3

Controlling Filamentous Fungal Morphology by Immobilization on a Rotating Fibrous Matrix to Enhance Oxygen Transfer and L(+)-Lactic Acid Production by *Rhizopus oryzae*

Nuttha Thongchul and Shang-Tian Yang*

Department of Chemical Engineering, The Ohio State University, 140 West 19th Avenue, Columbus, OH 43210

Filamentous fungi are widely used in industrial fermentations. However, the filamentous morphology is usually difficult to control and often cause problems in conventional submerged fermentations. The fungal morphology has profound effects on mass transfer, cell growth, and metabolite production. Controlling the filamentous morphology by immobilization on a rotating fibrous matrix was studied for its effects on oxygen transfer and lactic acid production in aerobic fermentation by *Rhizopus oryzae*. Compared to the conventional stirred tank fermentor, the fermentation carried out in the rotating fibrous bed bioreactor (RFBB) resulted in a good control of the filamentous morphology, and improved oxygen transfer and lactic acid production from glucose. A high lactic acid concentration of 137 g/L with a high yield of 0.83 g/g and reactor productivity of 2.1 g/L·h was obtained with the RFBB in repeated batch fermentations. The process was stable and can be used for an extended operation period.

Introduction

Lactic acid is commercially produced by either chemical synthesis or fermentation. In chemical synthesis, lactic acid is generally produced by the hydrolysis of lactonitrile formed in the reaction of acetaldehyde with hydrogen cyanide (*1*). *Lactobacillus sp.* have been commonly used in lactic acid fermentation due to their high volumetric productivity and yield (*2-4*). However, the racemic mixtures of L(+)- and D(-)-lactic acids produced by most *Lactobacillus sp.* are difficult to use in the manufacture of biodegradable polylactic acid (*5*). Although some mutants can produce pure L(+)-lactic acid, they are not commonly available. Furthermore, *Lactobacillus* requires complex media for growth (*6, 7*), which makes the final product recovery and purification difficult and costly. Recently, there have been increasing interests in fungal fermentation with *Rhizopus oryzae* to produce optically pure L(+)-lactic acid from glucose, pentose sugars and starch directly in a simple medium (*8-13*). However, it is cumbersome to control the filamentous morphology in conventional submerged fermentations (*14-16*), which greatly hampers reactor operation and limits the fungal fermentation process due to lowered product yield and production rate. Table I compares homolactic *Lactobacillus sp.* and *R. oryzae* in their use for lactic acid production.

Table I. Comparison between bacterial and fungal lactic acid fermentations.

	Lactobacillus sp.	*R. oryzae*
Substrates	can't use starch and pentoses	can use starch and pentoses
Medium	require complex growth nutrients	simple medium composition
Growth conditions	anaerobic, pH > 4.5	aerobic, pH > 3
Products	usually mixtures of L(+) and D(-)-lactic acids	pure L(+)-lactic acid, plus other byproducts (e.g., ethanol, fumarate, CO_2)
Product yield from glucose	0.85 ~ 0.95 g/g	usually less than ~0.85 g/g
Product concentration	up to 150 g/L	up to ~130 g/L
Productivity	can be as high as 60 g/L·h	usually lower than 6 g/L·h
Reactor operation	easy	difficult due to the filamentous cell morphology

Figure 1 shows a generalized catabolic pathway found in *R. oryzae*. *R. oryzae* usually converts glucose to pyruvic acid via the EMP pathway. *R. oryzae* can also use pentose phosphate pathway (HMP) in pentose sugar catabolism. In addition, *R. oryzae* has amylases and can convert starch to glucose. Oxygen is a critical factor affecting lactic acid production from pyruvate (*17*). In the presence of glucose and high dissolved oxygen concentration, pyruvic acid is converted to lactic acid by an NAD^+-dependent lactate dehydrogenase (LDH) (*17*). During glucose depletion or sporulation, the activity of LDH rapidly decreases and the oxidation of lactic acid to pyruvic acid is usually found. The metabolism of pyruvic acid in *R. oryzae* also involves the tricarboxylic acid (TCA) cycle occurring in mitochondria and the separated cytosolic pathway for fumaric acid formation involving pyruvate carboxylase, malate dehydrogenase and fumarase. Although *R. oryzae* does not grow anaerobically, it possesses alcohol dehydrogenase (ADH), which allows the fungus to grow in a short period in the absence of oxygen. Depending on the fermentation conditions, ethanol, CO_2 and fumaric acid also can be produced as major byproducts in addition to lactic acid by *R. oryzae*.

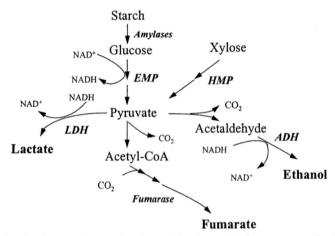

Figure 1. Pathways for production of lactic acid, fumarate, and ethanol in Rhizopus oryzae. (LDH: lactate dehydrogenase; ADH: alcohol dehydrogenase)

Control of the mycelial morphology and broth rheology is important to the fungal fermentation. The highly branched fungal mycelia may cause complex (viscous) broth rheology and difficulty in mixing and aeration in the conventional agitated tank fermentor (*14-16*). Various cell immobilization methods to control the cell morphology and to achieve high cell density and high reaction rate have been studied (*18-23*). In general, higher lactic acid yield and productivity were achieved with immobilized cells than those from free mycelial cells, partially due to reduced cell growth and increased specific cell productivity. With the immobilized cells, a high mycelial biomass density also can be achieved, which can be repeatedly used for lactic acid production over a long period. Power consumption in mixing and aeration also can be greatly reduced because the fermentation broth is maintained at a low viscosity (similar to water) under cell-free condition, which also facilitates product recovery from the relatively pure and simple medium by solvent extraction (*10*).

In this work, we immobilized fungal spores and mycelia in a rotating fibrous matrix in a stirred-tank bioreactor, and used the reactor to produce L(+)-lactic acid from glucose for an extended period of more than 10 days, demonstrating the feasibility and advantages of the new fungal bioreactor. The effects of cell immobilization on fungal morphology, oxygen transfer, and lactic acid production by *R. oryzae* were studied. The dynamic method of gassing out was used to evaluate oxygen transfer and uptake during the fermentation. The effects of aeration rate on the fungal fermentation and RFBB performance were also studied and the results are discussed in this paper.

Materials and Methods

Microorganism, Media, and Inoculum Preparation

The stock culture of *R. oryzae* NRRL 395 was maintained on potato dextrose agar (PDA) plates (*10*). For spore germination and initial cell growth, a growth medium containing 50 g/L glucose and 5 g/L yeast extract was used. For lactic acid production, the medium consisted of (per liter): 70 g glucose, 0.6 g KH_2PO_4, 0.25 g $MgSO_4$, 0.088 g $ZnSO_4$, and 0.3 g urea. Antifoam A was manually added to prevent foaming during fermentation. The sporangiospores were collected from PDA petri plates by shaving and extracting the spores with sterile water. Spore count was done and the number of spores in the suspension was adjusted to 10^6/mL by dilution with sterile water, and then used to inoculate the bioreactor.

Rotating Fibrous Bed Bioreactor (RFBB)

Figure 2 shows a schematic of the RFBB, which was modified from a 5-L fermentor (Biostat B, B. Braun) by affixing a perforated stainless steel cylinder covered with a cotton cloth (9 cm × 15 cm × 0.2 cm) to the agitation shaft. Before use, the bioreactor was sterilized twice at 121°C for 60 minutes with an overnight interval. The bioreactor containing 4 liters of the growth medium was then autoclaved at 121°C for 60 minutes. After cooling, the dissolved oxygen (DO) probe was calibrated with air and nitrogen, and the reactor was then inoculated with 10 mL of the spore suspension (10^6 spores/mL). Unless otherwise noted, the bioreactor was controlled at 30°C and pH 6.0, agitated at 50 rpm, and aerated with filter-sterilized air at 0.5-2.0 vvm.

To study the fermentation kinetics, the growth medium was replaced with the production medium after spore germination and cell immobilization on the fibrous matrix had occurred. To study the long-term stability of the bioreactor, repeated batch fermentations were conducted by replacing the medium at the end of each batch when glucose was almost depleted and lactic acid production stopped. A total of 9 batches were conducted consecutively with each batch taking about 24 h. Repeated batch fermentations with high-concentration glucose media were also conducted to evaluate the maximum lactic acid concentration that can be produced by the fungal cells. About 2 L of the medium in the bioreactor were replaced with a concentrated medium whenever glucose was almost depleted, until the fermentation finally stopped due to inhibition of the cells by high-concentration lactic acid.

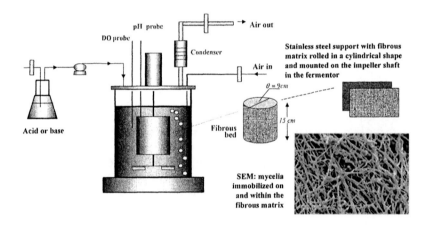

Figure 2. A rotating fibrous bed bioreactor (RFBB) used in this study for filamentous fungal fermentation. The scanning electron micrograph (SEM) shows fungal mycelia grew both inside and on the surface of the fibrous matrix.

Oxygen Transfer Experiments

Before the end of each batch fermentation, the oxygen uptake rate (OUR) and the dynamics of oxygen transfer in the reactor were studied using the dynamic method of gassing out. In the experiment, aeration was stopped to allow the dissolved oxygen concentration (C_L) in the medium to drop, and aeration was then resumed. Figure 3 shows a typical oxygen concentration profile during one experiment. The change in C_L with time (dC_L/dt) can be expressed by the following equation:

$$dC_L/dt = OTR - OUR \tag{1}$$

where OTR is the oxygen transfer rate in the medium due to aeration and can be expressed as:

$$OTR = k_La \cdot (C^* - C_L) \tag{2}$$

where C^* is the solubility of oxygen in the medium (mM) and k_La is the volumetric mass transfer coefficient (min⁻¹). When there is no aeration, OTR = 0 and OUR (mM/min) can be determined from the slope of the plot of C_L versus time during the period without aeration. Both k_La and C^* can then be estimated from the plot of (dC_L/dt + OUR) versus C_L during the period with aeration, with k_La equal to the negative slope and C^* equal to the Y-intercept divided by k_La.

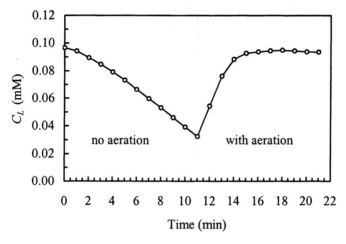

Figure 3. A typical dissolved oxygen profile in the bioreactor obtained using the dynamic method of gassing out in the oxygen transfer experiment.

Analytical Methods

The fermentation broth sample was centrifuged to remove the suspended solids and the supernatant was analyzed for glucose and L(+)-lactic acid using a glucose/lactate analyzer (YSI 2700, Yellow Spring, OH). High performance liquid chromatography (HPLC) was also used to analyze the organic compounds (glucose, lactic acid, fumaric acid, and ethanol) present in the fermentation broth. The HPLC system (Shimadzu Scientific Instruments) was equipped with a RID-10A reflective index detector and an organic acid analysis column (Aminex HPX-87H, BioRad). Cell dry weight was determined by drying in an oven at 105°C until constant weight was obtained.

Results and Discussion

Effect of Immobilization on Fungal Morphology

After spore germination, cells generally grew into large mycelial clumps that were clinged to surfaces of agitation shaft, probes, and fermentor wall in the conventional stirred-tank bioreactor (Figure 4A), which made the bioreactor difficult to operate and control. However, when cotton cloth was present for cell attachment and immobilization in the RFBB, fungal mycelia were only attached on the cotton cloth and no cells were found in the fermentation broth or any other surfaces (Figure 4B). Consequently, the fermentation broth was clear and the bioreactor was easy to operate and control. It is clear that cotton cloth provided a preferential surface for cell attachment and the shear acting upon the rotating mycelial layer helped maintaining its compactness. The different fungal morphologies found under these two different fermentation conditions also resulted in significant difference in lactic acid production by the cells. The fermentation kinetics obtained in these two bioreactor systems are shown in Figures 5A and 5B, respectively.

Figure 4. Fungal morphology seen in the fermentor without cotton cloth, where fungal mycelia forming large clumps attached everywhere in the fermentor (A), and in the RFBB, where fungal mycelia forming a sheet layer attached only on the cotton cloth in the fermentor (B).

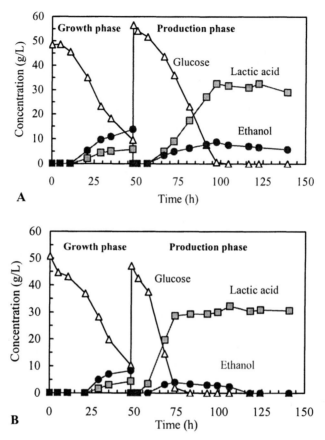

Figure 5. Fermentation kinetics during the growth phase and production phase by free cells in a stirred-tank fermentor (A) and by immobilized cells in the RFBB (B) at 50 rpm, 0.5 vvm, 30°C and pH 6.

Fermentation Kinetics

Figure 5A shows typical fermentation kinetics with fungal cells in a conventional stirred-tank fermentor. During the growth phase, the carbon source (glucose) was mainly used for growth and synthesis of cell biomass, and there were some ethanol and a smaller amount of lactic acid produced. In the production phase, cell growth was greatly reduced due to nitrogen limitation and glucose was mainly used to produce lactic acid with ethanol as a fermentation byproduct. Similar fermentation kinetics was observed with the RFBB (Figure 5B); however, faster and more lactic acid production was obtained with less ethanol produced in the production phase. It is well known that cell growth and lactic acid production by *R. oryzae* require oxygen (*17, 23*). More ethanol would be produced when there was oxygen limitation because the activity of lactate dehydrogenase (LDH) declined and alcohol dehydrogenase (ADH) was induced, resulting in the oxidation of lactic acid to pyruvic acid, which then entered the ethanol fermentative pathway. The higher lactic acid production rate and yield obtained in the RFBB were attributed to the better oxygen transfer as indicated by the higher oxygen uptake rate (OUR) and volumetric oxygen transfer coefficient (k_La) which were determined from the dynamic gassing out method.

Cell growth could not be easily measured for immobilized cell fermentation in the RFBB. The mycelial layer was ~0.5 cm thick from the surface of the cotton cloth at the end of the growth phase, and increased to ~1 cm at the end of the production phase. The nutrients present in yeast extract were essential to spore germination, but not critical to mycelial growth. It is also possible that the residual nutrients left in the fermentation broth from the growth phase were sufficient to support continued cell growth in the production phase. The cell density at the end of the production phase was estimated by harvesting the cells in the bioreactor and measuring the total cell dry weight. More cell biomass was found in the RFBB than in the free-cell fermentation, indicating faster growth in the RFBB due to improved aeration (higher C_L and k_La values). Table II summarizes and compares the experimental results from the RFBB and the conventional stirred-tank fermentor. It is clear that the RFBB gave better oxygen transfer and higher cell growth and lactic acid production, all of which can be attributed to the better controlled fungal morphology caused by cell immobilization on the rotating fibrous matrix.

Table II. Comparison of free-cell fermentation in the conventional stirred-tank fermentor and immobilized cell fermentation in the RFBB.

		Free cell	RFBB
Product yield (g/g)	Lactic acid	0.597	0.686
	Ethanol	0.133	0.074
Productivity (g/L·h)	Lactic acid	0.82	1.58
	Ethanol	0.20	0.14
Total cell dry weight (g)		10.97	18.62
Oxygen uptake rate (mM/min)		0.0086	0.010
C^* (mM)		0.17	0.17
C_L (mM)		0.09	0.13
$k_L a$ (min^{-1})		0.13	0.36

Note: fermentation conditions: 50 rpm, 0.5 vvm, 30°C and pH 6.

It should be noted that the values of k_La and C^* during the fermentation were estimated from the plot of $(dC_L/dt + OUR)$ versus C_L using the data obtained in the dynamic gassing out experiments conducted right after the fermentation was completed. The estimated C^* value of 0.17 mM was significantly lower than the solubility of oxygen in water (~0.24 mM for air at 1 atm) at 30°C (24). The lower C^* value could be partially attributed to the electrolytes present in the medium and CO_2 production during the fermentation.

Long-Term Lactic Acid Production in the RFBB

The feasibility and stability of the fungal culture immobilized in the RFBB for long-term production of lactic acid from glucose were evaluated in repeated batch fermentations operated at 50 rpm, 1.0 vvm, pH 6.0 and 30°C. As shown in Figure 6A, except for the initial batch for cell growth, all subsequent 9 batches gave consistent lactic acid production during the 10-day period studied. There was no ethanol or other byproducts found in the fermentation at the increased aeration rate of 1 vvm. The lactic acid yield and volumetric productivity from each batch fermentation are shown in Figure 6B, along with the OUR and k_La values estimated from the dynamic gassing out experiments performed at the end of each batch. As can be seen in Figure 6B, the lactic acid yield increased steadily from 0.53 to 0.74 g/g, while the reactor productivity remained relatively unchanged at 1.6 ± 0.1 g/L·h. In the mean time, the oxygen transfer coefficient k_La remained almost constant at ~0.44 ± 0.03 min⁻¹, C_L generally increased from ~0.11 mM in the first batch to ~0.15 mM in the 9th batch, while OUR fluctuated from batch to batch around 0.01 mM/min. The increased product yield could have been due to the increased dissolved oxygen concentration in the medium. However, it should be noted that the oxygen transfer experiments had relatively large experimental errors due to the instability of the dissolved oxygen probe during the long study period.

The increasing lactic acid yield during the repeated batch fermentations also could have been due in part to the reduced cell growth. Although there was continued cell growth as evidenced by the increasing thickness of the mycelial layer on the rotating fibrous matrix, cell growth reduced significantly in the successive batch fermentations. The thickness of the mycelial layer on the rotating fibrous matrix increased from 0.5 cm to 1 cm during the first production batch, but only increased to ~2 cm at the end of the 9th batch. The reduced cell growth might have allowed more glucose for lactic acid production.

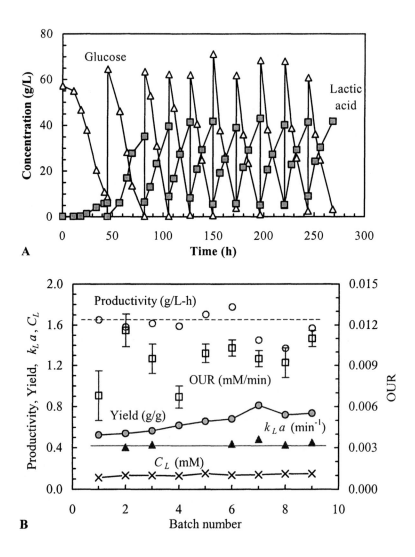

Figure 6. Kinetics of repeated batch fermentations in the RFBB at 50 rpm, 1.0 vvm, 30°C and pH 6. The first batch was for cell growth in the growth medium, and all subsequent batches were conducted with the production medium. (A) Glucose and lactate concentration profiles; (B) Yield, productivity and oxygen transfer data.

Since oxygen transfer is critical to lactic acid production, further study of the fermentation was conducted with the RFBB operated at a higher aeration rate of 2.0 vvm. The fermentation was also operated at repeated batch mode with partial medium replacement at the end of each production batch. In order to evaluate the maximum lactic acid concentration that can be produced by the fermentation, high-concentration glucose media were used in the last three batches. As shown in Figure 7A, except for the initial batch for cell growth, lactic acid production was stable during the 11-day period studied and there was no ethanol or other byproducts formed in the fermentation. The highest lactic acid concentration reached in the last batch fermentation was 137.3 g/L, which is the highest ever reported in the literature (*23*). Figure 7B shows the lactic acid yield and volumetric productivity from each of the 8 production batches, along with the estimated OUR and k_La values from the dynamic gassing out experiments. It is clear that with the increased aeration rate of 2 vvm, the dissolved oxygen concentration (C_L) in the fermentor increased to ~0.15 mM due to improved oxygen transfer with a high $k_L a$ of 0.67 ± 0.02 min^{-1}. Consequently, the fermentation gave a consistently higher lactic acid yield of ~0.83 ± 0.06 g/g. The reactor productivity was also higher, reaching ~2.1 g/L· h in the last four batches even though the lactic acid concentration was high. OUR was significantly lower in the last two batches, perhaps due to inhibition by lactic acid and/or glucose at high concentrations.

Effects of Oxygen on Fermentation

It should be noted that both lactic acid production rate and yield in the RFBB can be further improved by improving oxygen transfer. Unlike homolactic acid bacteria, *R. oryzae* requires oxygen for growth and lactic acid production. The dissolved oxygen concentration was found to be a critical factor affecting lactic acid production from glucose (*23*). A high dissolved oxygen concentration was desirable for lactic acid production. At low dissolved oxygen concentrations, the fermentation was not only slower but also produced less lactic acid with more ethanol as the byproduct due to increased activity of alcohol dehydrogenase (*17*). As can be seen in Table III, the reactor volumetric productivity and lactic acid yield from glucose increased with increasing the aeration rate due to increased dissolved oxygen concentration C_L resulted from more efficient oxygen transfer as indicated by the higher k_La value in the reactor. Also, the RFBB had better oxygen transfer and lactic acid production than the free-cell fermentation at the same aeration rate. High dissolved oxygen concentration was necessary for efficient oxygen diffusion into the relatively thick mycelial layer in the RFBB or the large mycelial clumps seen in the stirred tank fermentor.

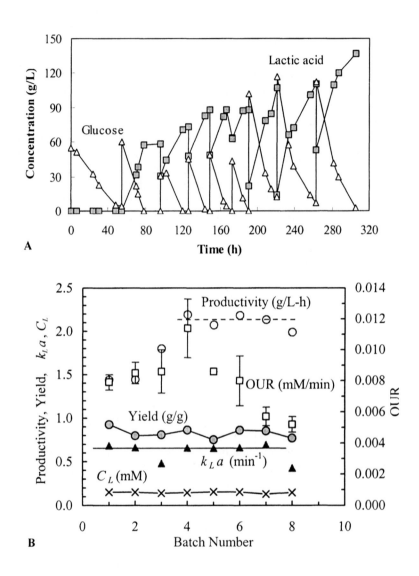

Figure 7. Kinetics of repeated batch fermentations in the RFBB at 50 rpm, 2.0 vvm, 30°C and pH 6. The first batch was for cell growth in the growth medium, and all subsequent batches were conducted with the production medium. About half of the fermentation broth was replaced with fresh media containing high concentrations of glucose at the end of each batch. (A) Glucose and lactate concentration profiles; (B) Yield, productivity and oxygen transfer data.

Table III. Effects of aeration on lactic acid fermentations by free cells and immobilized cells in the RFBB.

Morphology	Aeration rate (vvm)	k_La (min^{-1})	C_L (mM)	Yield (g/g)	Productivity (g/L·h)
Free cells	0.5	0.13	0.09	0.60	0.82
Immobilized	0.5	0.36	0.13	0.69	1.58
cells in RFBB	1.0	0.44±0.03	0.14±0.01	0.72±0.06	1.60±0.10
	2.0	0.67±0.02	0.15±0.01	0.83±0.06	2.14±0.05

As already discussed earlier, during the repeated batch fermentations the fungal cells continued to grow and the mycelial layer on the rotating fibrous matrix continued to increase to ~2 cm. However, the lactic acid production rate did not increase with increasing fungal cell biomass. Diffusion limitation and oxygen starvation must have occurred to the inner fungal cells that were away from the surface of the mycelial layer, resulting in lower specific productivity for cells inside the mycelial layer. Diffusion limitation can be alleviated by increasing the dissolved oxygen concentration in the medium, which can be achieved by either increasing the oxygen solubility in the medium or increasing the oxygen transfer rate. The former can be achieved by applying a higher air pressure or using pure oxygen to increase the partial pressure of oxygen in the air; the latter can be done by increasing agitation and aeration rates. Lactic acid yield could be increased to more than 0.9 g/g when the dissolved oxygen concentration was increased by using oxygen-enriched air (*23*).

It is well known that filamentous fungi could have different morphologies under different fermentation conditions, particularly due to differences in oxygen tension and mechanical force or shear (*25, 26*). The filamentous fungal morphology often causes difficulties in agitation and aeration because of its complicated effects on the rheological property of the fermentation broth and reactor hydrodynamics. This morphological problem has been alleviated by controlling fungal growth into small pellets (*13, 16*) or immobilization in porous materials by entrapment (*18, 19, 27*) or on solid surfaces by attachment (*20*). However, long-term performance of these fungal fermentation systems have not been well studied (*27, 28*), and further improvements in lactic acid yield and production rate are needed in order to compete with the commonly used bacterial fermentation process. With the RFBB, *R. oryzae* gave relatively stable, high production rate and yield for an extended period. However, the reactor productivity would be limited by the available surface area for cell attachment and oxygen transfer by diffusion into the mycelial layer. Further studies to increase the surface area in the RFBB and to control mycelial layer thickness are needed before the reactor can be scaled up for commercial applications.

50

Conclusion

This study demonstrated the feasibility and advantage of using the RFBB to control the filamentous fungal morphology in lactic acid production by *R. oryzae*. The cotton cloth used in the RFBB provided a preferential matrix for immobilizing fungal spores and mycelia, resulting in a cell-free broth that was better for fermentation operation and control. The improved oxygen transfer in the RFBB not only increased the fermentation rate and lactic acid production, but also eliminated undesirable byproduct ethanol and allowed the bioreactor to be used for long-term production. Since the dissolved oxygen concentration is a critical factor affecting lactic acid production, methods to enhance oxygen transfer into the mycelial layer should be further investigated.

Acknowledgements

This work was supported in part by research grants from US Department of Agriculture and Midwest Advanced Food Manufacturing Alliance.

References

1. Benninga, H. A history of lactic acid making. Kluwer Academic. **1990**.
2. Litchfield, J.H. *Adv. Appl. Microbiol.* **1996**, 42, 45-95.
3. Hofvendahl, K.; Hahn-Hagerdal, B. *Enz. Microb. Technol.* **2000**, 26, 87-107.
4. Vick Roy, T.B. Lactic acid. In: Moo-Young M, editor. Comprehensive Biotechnology. Oxford: Pergamon. **1985**. pp. 761-776.
5. Tsai, S.P.; Moon, S.-H. *Appl. Biochem. Biotechnol.* **1998**, 70-72, 417-428.
6. Hsieh, C.M.; Yang, F.-C.; Iannotti, E.L. *Process Biochem.* **1999**, 34, 173-179.
7. Silva, E.M; Yang, S.T. *J. Biotechnol.* **1995**, 41, 59-70.
8. Ho, W. Kinetics of L(+)-lactic acid production from glucose, xylose, and starch by free cells and immobilized cells of *Rhizopus oryzae*. MS Thesis, Ohio State University, Columbus, Ohio. 1996.
9. Soccol, C.R.; Stonoga, V.I.; Raimbault, M. *World J. Microbiol. Biotechnol.* **1994**, 10, 433-435.
10. Tay, A. Production of *L(+)*-lactic acid from glucose and starch by immobilized cells of *Rhizopus oryzae*. PhD thesis. Ohio State University. Columbus, Ohio. 2000.
11. Woiciechowski, A.L.; Soccol, C.R.; Ramos, L.P.; Pandey, A. *Process Biochem.* **1999**, 34,949-955.
12. Yu, R.; Hang, Y.D. *Biotechnol Lett.* **1989**, 11, 597-600.
13. Zhou, Y.; Dominguez, J.M.; Cao, N.; Du, J.; Tsao, G.T. *Appl. Biochem. Biotechnol.* **1999**, 77-79, 401-407.

14. Kosakai, Y.; Park, Y.S.; Okabe, M. *Biotechnol. Bioeng.* 1997, 55, 461-470.
15. Park, E.Y.; Kosakai, Y.; Okabe, M. *Biotechnol Progr.* **1998**, 14, 699-704.
16. Yang, C.W.; Lu, Z.; Tsao, G.T. *Appl. Biochem. Biotechnol.* **1995**, 51-52, 57-71.
17. Skory, C.D.; Freer, S.N.; Bothast, R.J. *Biotechnol. Lett.* **1998**, 20, 191-194.
18. Dong, X.Y.; Bai, S.; Sun, Y. *Biotechnol. Lett.* 1996, 18, 225-228.
19. Hang, Y.; Hamamci, H.; Woodams, E.E. *Biotechnol. Lett.* **1989**, 11, 119-120.
20. Lin, J.P.; Ruan, S.D.; Cen, P.L. *Chem. Eng. Commun.* **1998**, 168, 59-79.
21. Sun, Y.; Li, Y.-L.; Bai, S. *Biochem. Eng. J.* **1999**, 31, 87-90.
22. Tamada, M.; Begum, A.A.; Sadi, S. *J. Fermentation Bioeng.* **1992**, 74, 379-383.
23. Tay, A.; Yang, S.T. *Biotechnol. Bioeng.* **2002**, 80, 1-12.
24. Bailey, J.E.; Ollis, D.F. *Biochemical Engineering Fundamentals;* McGraw-Hill, New York, 1986; 2nd ed., pp. 457-507.
25. Cui, Y.Q.; Van der Lans, R.G.J.M.; Luyben, K.C.A.M. *Biotechnol. Bioeng.* **1998**, 57, 409-419.
26. Cui, Y.Q.; Okkerse, W.J.; Van der Lans, R.G.J.M.; Luyben, K.C.A.M. *Biotechnol. Bioeng.* **1998**, 60, 216-229.
27. Yin, P.; Yahiro, K.; Ishigaki, T.; Park, Y.; Okabe, M. *J. Ferment. Bioeng.* **1998**, 85, 96-100.
28. Sun, Y.; Li, Y.-L.; Bai, S.; Yang, H.; Hu, Z.-D. *Bioprocess Eng.* **1998**, 19, 155-157.

Chapter 4

Enhancing Butyric Acid Production with Mutants of *Clostridium tyrobutyricum* Obtained from Metabolic Engineering and Adaptation in a Fibrous-Bed Bioreactor

Ying Zhu and Shang-Tian Yang*

Department of Chemical Engineering, The Ohio State University, 140 West 19th Avenue, Columbus, OH 43210

Butyric acid has wide applications in food and pharmaceutical industries. Its production by fermentation from natural resources has become an increasingly attractive alternative to the petroleum-based production route currently used in the chemical industry. In this work, novel metabolic engineering approaches, at both molecular biology and process engineering levels, were developed for enhanced butyric acid production by *Clostridium tyrobutyricum*. Recombinant DNA technology was used to knock out genes in the acetate formation pathway and to overexpress genes in the butyrate formation pathway in mutant strains with improved butyrate production as compared to the wild-type strain. Also, a novel fibrous bed bioreactor (FBB) was used for fermentation of xylose to produce butyrate with enhanced reactor productivity, product concentration and yield. Cells in the FBB were able to grow into high density and adapt to tolerate a higher butyrate concentration, which was not achievable in conventional fermentation systems.

Introduction

Plant biomass is the only foreseeable sustainable source of organic fuels, chemicals, and materials. Recently, increasing concerns about future scarcity, cost, and environmental impact of fossil fuel have enlarged public interest in the technologies for using cheap renewable biomass. Biologically based processing technologies provide an attractive route to develop more energy efficient and environmentally sustainable methods for producing a wide range of products, including fuels, chemicals, plastics, pharmaceuticals, and industrial solvents. Biomass feedstocks include agricultural residues and industrial wastes, which usually require costly disposal to avoid pollution problems. Successful conversion of biomass to products requires an efficient bioprocess for economically viable industrial applications. In this work, we aimed at developing a novel fermentation process to economically produce butyric acid from low-value agricultural commodities such as corn and its byproducts which are rich in both glucose and xylose.

Butyric acid has many applications in the chemical industry as well as food and pharmaceutical industries. It is used in the form of pure acid to enhance butter-like notes in food flavors. Esters of butyric acid are used as additives for increasing fruit fragrance and as aromatic compounds for production of perfumes (1). Butyric acid is one of the short-chain fatty acids generated by microbial fermentation of dietary substrates, and is considered to have therapeutic nature for the treatment of colorectal cancer and hemoglobinopathies (2). Drugs derived from butyric acid have been widely studied and developed. Butyric acid is currently produced mainly by oxidation of butyraldehyde obtained from propylene by oxosynthesis, with a market price of $1.21/kg (3). However, the demand for butyric acid from microbial fermentation is high due to increasing health concerns and a strong interest in using biologically produced food additives preferred by food manufacturers. The potential markets for bio-based butyric acid and its esters are thus big and awaiting for exploration.

However, current technology for bio-production of butyric acid is not competitive as compared with petrochemical production because butyric acid-producing bacteria convert sugars to acetic acid in addition to butyric acid as their major fermentation products (4) and are inhibited by butyric acid (5). Conventional fermentation processes usually suffer from low final product concentration, low reactor productivity, and low product yield. A combination of classical genetics, bioprocess engineering, and metabolic manipulation can be used effectively to improve biosynthetic processes (6). In this study, we worked on both molecular biology and process engineering levels to develop an efficient bioprocess for butyric acid production from glucose and xylose.

The main objective of this study was to obtain butyrate-producing mutants of *Clostridium tyrobutyricum* that are capable of producing butyrate with a high yield and tolerating a high butyrate concentration. Genetic improvements in microbial cultures have been made to channel metabolic intermediates specifically toward a desired product (7). Since there usually is concomitant production of acetate in butyric acid fermentation, disruption of the genes *ack* and *pta* involved in the acetate formation pathway should improve the butyrate yield from sugars fermented. Meanwhile, butyrate production may be further enhanced by overexpressing genes (*buk* and *ptb* genes) in the butyrate formation pathway. In this work, we have developed mutant strains with improved characteristics for butyric acid fermentation by gene inactivation and overexpression techniques. The effects of gene disruption and overexpression on cell growth and fermentation kinetics of the mutants were studied and are reported here. To achieve a high butyrate concentration in the fermentation, we used a fibrous bed bioreactor (FBB) previously developed for organic acid fermentations (8-12) to adapt the cells to tolerate a higher butyrate concentration. The feasibility and advantages of the FBB for butyric acid production from xylose were also evaluated in this study. It was found that with high densities of cells immobilized in the fibrous matrix, the FBB greatly increased reactor productivity, final product concentration, and product yield as compared with conventional free-cell fermentations.

Materials and Methods

Culture and Media

The bacterium *C. tyrobutyricum* ATCC 25755 was used for butyric acid fermentation. It was cultured in a clostridial growth medium (CGM) described previously (9) with either glucose or xylose as the substrate. The stock culture was kept in serum bottles under anaerobic conditions at 4°C. In the molecular biology study, *C. tyrobutyricum* was grown anaerobically at 37°C in Reinforced Clostridial Medium (RCM, Difco). Colonies were maintained on RCM plates. RCM or CGM medium was supplemented, as required, with 40 µg/ml erythromycin (Em) or 20 µg/ml thiamphenicol (Th). *E. coli* used in the cloning work was grown aerobically at 37°C in Luria-Bertani (LB) medium supplemented with ampicillin (100 µg/ml) and erythromycin (200 µg/ml).

Mutant Development by Genetic Engineering

DNA Isolation and Manipulation. Isolation of plasmid DNA from *E. coli* was undertaken using QIAprep Miniprep plasmid purification kit (Qiagen). Restriction enzymes, T4 ligase, and shrimp alkaline phosphatase were used in accordance with the supplier's instruction (Amersham Pharmacia). Genomic DNA from *C. tyrobutyricum* was isolated by using QIAGEN genomic DNA kit.

PCR Amplification. Several synthetic oligonucleotides (Integrated DNA) were designed and used as primers for PCR. The sequences of the PCR primers for *ack* gene were 5'– GAT AC(A/T) GC(A/T) TT(C/T) CA(C/T) CA(A/G) AC –3' and 5'– (G/C)(A/T)(A/G) TT(C/T) TC(A/T) CC(A/T) AT(A/T) CC(A/T) CC –3'. The sequences of primers for *pta* gene were 5'– GA(A/G) (C/T)T(A/T/G) AG(A/G) AA(A/G) CA(T/C) AA(A/G) GG(A/T) ATG AC– 3' and 5'–(A/T)GC CTG (A/T)(G/A)C (A/T)GC(A/T/C) GT(A/T) AT(A/T) GC– 3'. Thermal cycling was performed to carry out the amplification in a DNA engine (MJ Research), using *C. tyrobutyricum* genomic DNA as the template. DNA fragments of *ack* gene and *pta* gene with expected sizes of 560 bp and 730 bp respectively, were amplified, and then cloned into PCR vector pCR 2.1 to form pCR-AK and pCR-PTA using TA cloning (Invitrogen).

Construction of Integrational Plasmids. A 1.5 kb Sph I fragment was removed from pCR-AK (4.6 kb) and pCR-PTA (4.75 kb), and the vectors were religated to form pCR-AK1 and pCR-PTA1. A 1.6 kb *Hind*III fragment containing the Em^r cassette was removed from pDG 647 (13), and then ligated into *Hind*III digested pCR-AK1 and pCR-PTA1 to form the integrational plasmids pAK-Em (4.7 kb) and pPTA-Em (4.85 kb).

Construction of Replicative Plasmids. The butyrate operon (*ptb* and *buk* genes) from *C. acetobutylicum* was amplified by PCR using plasmid pJC7 as template, subcloned into pCR 2.1, and then digested by *EcoR*I. A 2.2 kb *EcoR*I fragment containing butyrate operon was subcloned into pIMPTH to form pTHBUT (7.1 kb) (14).

Transformation. Plasmid transformation to *E. coli* was performed according to the manufacturer's instruction (Invitrogen). Transformation of plasmids into *C. tyrobutyricum* was carried out using a Bio-Rad Gene pulser in an anaerobic chamber. The competent cells were prepared as follows: mid exponential-growth phase cells grown in CGM were harvested, washed twice and suspended in ice-cold electroporation buffer (SMP; 270 mM sucrose, 7 mM sodium phosphate, pH 7.4, 1 mM MgCl$_2$). Cell suspension (0.5 ml) was chilled on ice for 5 min in a 0.4 cm electroporation cuvette (Bio-Rad), and plasmid DNA (1 µg) was added to the suspension and mixed well. After the pulse had been applied (2.5 kV, 600 Ω, 25 µF), cells were transferred to 5 ml RCM and incubated for 4 h at 37°C prior to plating on RCM containing 40 µg/ml Em or 20 µg/ml Th. The mutant strains containing pAK-Em, pPTA-Em, and pTHBUT were selected and are denoted as PAK-EM, PPTA-EM, and WT(pTHBUT), respectively.

Fermentation Kinetic Study. Fed-batch fermentations were carried out in 5-L stirred-tank fermentors to study the fermentation kinetics of various mutant strains and to evaluate the maximum butyric acid concentration achievable in fermentation.

Fibrous-Bed Bioreactor

Construction and Operation. The fibrous bed bioreactor was made of a glass column packed with spiral wound cotton towel and had a working volume of ~480 ml. Detailed description of the reactor construction has been given elsewhere (10). The reactor was connected to a 5-L stirred-tank fermentor (Marubishi MD-300) through a recirculation loop and operated under well-mixed condition with pH and temperature controls. Anaerobiosis was maintained by initially sparging the medium in the fermentor with N_2 and then kept the fermentor headspace under 5 psig N_2 during the entire fermentation run. Unless otherwise noted, the reactor containing 2 L of medium was maintained at 37°C, agitated at 150 rpm, and pH controlled at 6.0 by adding NH_4OH.

Fermentation and Culture Adaptation. To start the fermentation, ~100 ml of cell suspension in serum bottles were inoculated to the fermentor and allowed to grow for 3 days until the cell concentration reached an optical density (OD_{620nm}) of ~4.0. Cell immobilization was then carried out by circulating the fermentation broth through the fibrous bed at a pumping rate of ~25 ml/min to allow cells attach and be immobilized onto the fibrous matrix. After about 36~48 h of continuous circulation, most of the cells were immobilized and no change in cell density in the medium could be identified. The medium circulation rate was then increased to ~100 ml/min and the reactor was operated at a repeated batch mode to increase the cell density in the fibrous bed to a stable, high level (>50 g/L). To adapt the culture to tolerate a higher butyrate concentration, the reactor was then operated at fed-batch mode by pulse feeding concentrated substrate solution whenever the sugar level in the fermentation broth was close to zero. The feeding was continued until the fermentation ceased to produce butyrate due to product inhibition. Samples were taken at regular intervals for the analysis of cell, substrate and product concentrations. At the end of the fed-batch experiment, immobilized cells in the FBB were washed off from the fibrous matrix and stored at 4°C for further characterization.

Preparation of Cell Extracts and Acid-Forming Enzyme Assays

C. tyrobutyricum was grown in CGM (50 ml) at 37°C to the exponential phase (OD_{620} = ~1.5). Cells were harvested, washed and suspended in 25 mM Tris/HCl (pH 7.4). The cell suspension was sonicated, and cell debris was removed by centrifugation. The protein content of extracts was determined by the method of Bradford with bovine serum albumin as the standard (Bio-Rad protein assay). The activities of acetate kinase (AK) and butyrate kinase (BK) were measured in the direction of acyl phosphate formation based on the protocol of Rose (15). One unit of activity is defined as the amount of enzyme that produces 1 μmol of hydroxamic acid per minute under these conditions. Phosphotransacetylase (PTA) and phosphotransbutyrylase (PTB) were assayed by the method of Andersch et al. (16). One unit of enzyme is defined as the amount of enzyme converting 1 μmol of acyl-CoA or butyryl-CoA per minute under the reaction conditions. Specific activity of enzymes is defined as the units of activity per mg of protein.

Butyrate Tolerance Study

Cultures were grown in serum tubes containing 10 ml of media with various amounts of butyrate (0 – 15 g/L) to evaluate the inhibition effect of butyrate on cell growth, which was followed by measuring the optical density at 620 nm with a spectrophotometer. Specific growth rates were calculated from the growth data in the exponential phase.

Analytical Methods

Cell density was analyzed by measuring the optical density of cell suspension at 620 nm (OD_{620}) with a spectrophotometer. An HPLC system (Shimadzu) was used to analyze the organic compounds, including butyrate, acetate, glucose and xylose, present in the fermentation broth. Gas production was monitored using an on-line respirometer Micro-oxymax system equipped with both H_2 and CO_2 sensors (Columbus Instrument).

Results and Discussion

Gene Inactivation and Overexpression

Selective inactivation of genes on the chromosome using non-replicative integrational plasmids has emerged as a new genetic engineering technology to obtain mutants with desirable metabolic properties (17). It has been applied in the inactivation of several genes in *C. acetobutylicum* (18-20). The non-replicative integrational plasmid usually contains a DNA segment from a host in which they cannot replicate, and a genetic marker for which selection can be made. After transfer, the plasmid becomes established by inserting into the homologous regions on the host chromosome from which the DNA segment was derived. Integration occurs in a Campbell-like fashion (21), and results in duplicated homologous regions flanking the plasmid DNA. If the homologous DNA fragments are internal to the transcription unit, it will result in disruption of the unit and loss of function, possibly producing a mutant phenotype. This is called integrational mutagenesis. In this work, non-replicative plasmids pAK-Em and pPTA-Em were constructed and used to transform *C. tyrobutyricum* to disrupt the acetate-forming genes *ack* and *pta* (encoding AK and PTA in acetate formation) on the chromosome. Since the homologous regions in pAK-Em and pPTA-Em are internal genes of *ack* and *pta*, the insertion would be mutagenic and the original genes on the chromosome would be disrupted.

Exponential-phase cultures of *C. tyrobutyricum* wild-type, PAK-Em, and PPTA-Em were harvested and cell extracts were assayed for acetate and butyrate-producing enzymes (AK, PTA, BK, PTB). Strain PAK-Em displayed 54% lower AK activity and approximately 130% higher PTA activity than the wild-type. Strain PPTA-Em had only 20~40% of AK and PTA activities compared to the wild-type strain. These results indicated that *ack* was inactivated in mutant PAK-Em and *pta* was inactivated in mutant PPTA-Em. PPTA-Em also had reduced AK activity, indicating that the expression of *ack* was also inhibited. Similar result has been reported for *pta* deleted mutant of *C. acetobutylicum* (20). It has been reported that the acetate-forming genes *ack* and *pta* in *C. acetobutylicum* exist in the same operon on the chromosome with *pta* preceding *ack* (22). Likewise, a similar structure of acetate-forming genes with *ack* lying downstream from *pta* in the same operon may be present in *C. tyrobutyricum*.

Plasmid pTHBUT containing the butyrate operon from *C. acetobutylicum* was introduced into the wild-type *C. tyrobutyricum*. The presence of butyrate operon in pTHBUT resulted in 2~3 folds increase in the PTB activity and 30% reduction in AK and 50% reduction in PTA activities as compared with the wild-type strain. However, the BK activity was not affected at all. In contrast, the activities of PTB and BK increased by more than 6 folds in *C. acetobutylicum* (23, 24) and there was 2-fold increase in PTB and 40-fold increase in BK in the BK-deleted mutant of *C. acetobutylicum* (14) after introducing the overexpression plasmid. Apparently, the *buk* and *ptb* genes from *C. acetobutylicum* did not work as well in *C. tyrobutyricum*.

Culture Adaptation in FBB

Fed-batch fermentation was performed to adapt the wild-type strain to higher butyrate concentrations. After several fed-batches, cells in the FBB were removed and grown as suspension culture to examine their enzyme activities. The results were compared with those of the original culture used to seed the bioreactor. Based on the acid-forming enzyme assays, the adapted culture from the FBB showed ~65% higher PTB and ~50% higher BK activities, both of which are involved in butyrate production. Also, its PTB was ~18% less sensitive to butyrate inhibition than that from the original culture. The higher butyrate-forming enzyme activities must have contributed to faster butyrate production in the FBB culture observed in the fermentation study. These results indicate that the adapted culture from the FBB had a different phenotype from the original culture.

Fermentation Kinetic Studies

Fermentation studies were carried out in both genetically engineered mutants and wild-type (Figure 1). Both PAK-Em and PPTA-Em strains grew exponentially in the first two batches and then entered the stationary phase (Figure 1b and 1c). Acetate was produced after the lag phase and reached the maximum value of ~12 g/L after the third fed-batch. Butyrate concentration continued to increase until at the end of the fermentation due to product inhibition. The maximum level of butyric acid was 42~43 g/L in both *ack* and *pta* deleted mutant fermentations, which was 33% higher than that obtained in fermentation with the wild-type strain (Figure 1a). Table I summarizes the kinetic data for these fermentations. It is clear that *ack* and *pta* deletion gave lower acetate yield, higher butyrate productivity and final concentration, and consequently, higher selectivity of butyrate over acetate. The *ack* deleted mutant PAK-Em increased butyrate yield to 36% from 31% of the wild-type. The *pta* deleted mutant PPTA-Em had a similar butyrate yield as the wild-type but lower gas production.

Gene integration had a significant effect on cell growth. Both mutant strains, PAK-Em and PPTA-Em, have reduced growth rate compared to the wild-type, but reached higher biomass concentrations (Table I). This observation is similar to that found in *pta* or *buk* inactivated *C. acetobutylicum* (20). Since both acid-formation pathways are responsible for generating energy (ATP) for cells, a reduced acetate production may impose a metabolic burden on cells. A feasible cellular response to this metabolic burden is the elevation of the flux through the alternate ATP-generation pathway, namely butyrate formation, to avoid any significant loss in overall cell growth.

Compared to the wild-type strain, the overexpression mutant WT(pTHBUT) also grew slower with a specific growth rate of 0.196 h^{-1} (see Table I). As expected, WT(pTHBUT) had a higher butyrate productivity (0.52 vs. 0.48 g/L·h in wild-type) and higher butyrate yield (0.36 vs. 0.26 g/g glucose in wild-type) in the first batch. Acetate yield was very similar between WT(pTHBUT) (0.13 g/g glucose) and wild-type (0.11 g/g glucose). As a result, there was a greater proportion of C4 versus C2 derived products in WT(pTHBUT), and consequently, the selectivity of butyrate over acetate was improved. However, the final concentration of butyrate produced by WT(pTHBUT) was unexpectedly lower (~12 g/L) than the wild-type. As can be seen in Figure 1d, after the first batch, the cells stopped producing acids and cell density started to decrease significantly, indicating severe inhibition by butyrate, which will be further discussed later. An increase in butyrate production and decrease in acetate produdtion have been observed for the BK-deleted mutant of *C.*

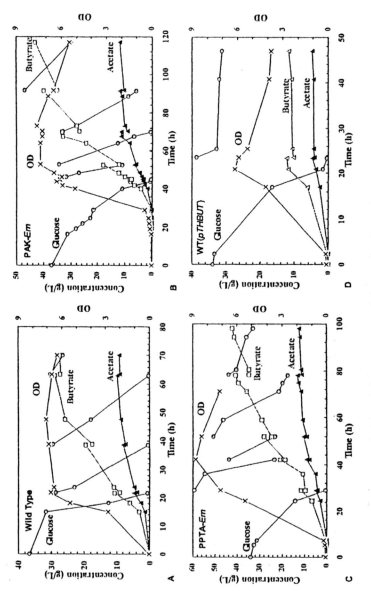

Figure 1. Fed-batch fermentation of glucose by C. tyrobutyricum at pH 6.0 and 37°C. (a) wild-type; (b) PAK-Em; (c) PPTA-Em; (d) WT(pTHBUT).

acetobutylicum transformed with the overexpression plasmid in the acidogenic phase (14); however, the final butyrate concentration in the solventogenic cultue of wild-type *C. acetobutylicum* harboring the overexpression plasmid also decreased (23). Gene overexpression in this case did not improve the butyrate fermentation, probably because the heterologous genes (enzymes) from *C. acetobutylicum* were too sensitive to butyrate inhibition and tended to shift metabolic pathway towards butyrate reassimilation (23). Further study should be carried out to overexpress the native *buk* and *ptb* genes.

Table I. Fermentation Characteristics of various *C. tyrobutyricum* strains in Fed-Batch Cultures Controlled at pH 6.0, 37 °C.

	Performance	*Wild-type*	*PAK-Em*	*PPTA-Em*	*WT (pTHBUT)*
Cell	Max. OD	7.1	7.6	8.8	5.9
	μ (h^{-1})	0.2762	0.1941	0.1668	0.196
	Yield (g/g)	0.128	0.113	0.147	0.120
Butyrate	Concentration (g/L)	28.6	43.0	42.1	12.2
	Yield (g/g)	0.315	0.360	0.310	0.360
	Productivity (g/L·h)	0.48~0.63	0.62~1.35	0.61~0.96	0.52
Acetate	Concentration (g/L)	9.7	11.9	12.2	4.8
	Yield (g/g)	0.106	0.098	0.080	0.132
Gas	H_2 yield (g/g)	0.015	0.032	0.012	0.005
	CO_2 yield (g/g)	0.305	0.494	0.251	0.084

Figure 2 shows the fermentation kinetics for butyric acid production from xylose by wild-type free cells and immobilized cells in the FBB. As compared to the free-cell fermentation, the immobilized-cell fermentation not only was faster but also produced a much higher butyrate concentration. The highest butyric acid concentration produced in the free-cell fermentation was only ~19 g/L, whereas butyric acid reached a concentration of ~58 g/L in the immobilized-cell fermentation with a yield of 0.47 g/g and a reactor productivity of 2.7 g/L·h. In contrast, the highest butyrate productivity in the free-cell fermentation was only 0.27 g/L·h. So far, the highest concentration of butyric acid obtained in fermentation was 62.8 g/L with a yield of 0.45 g/g sucrose by *C. tyrobutyricum*, but the reactor productivity was only 1.25 g/L·h (25). A much higher reactor productivity of 9.5 g/L·h at a butyrate concentration of 29.7 g/L was achieved for a continuous fermentation system with cell recycle by microfiltration (26). However, using a membrane filter for cell recycle to achieve a high cell density and reactor productivity could be a problem for long-term operation and process scale-up because dead cells would accumulate and foul the membrane, reducing system performance with time. The FBB used in this study gave good long-term performance without suffering from any fouling or other operational problems. The FBB also can be used to convert corn fiber hydrolysate and corn steep liquor, the byproducts from the corn-milling industry, to butyrate at a significantly lower cost (27). It is noted that there was a high density of cells (~70 g/L) immobilized in the fibrous matrix, which attributed to the higher fermentation rate, but the immobilized cells probably did not grow as much as the free cells suspended in the medium, indicated by the lower OD. Therefore, more substrates were converted to final products in the immobilized-cell fermentation. These results clearly indicate that the adapted culture immobilized in the FBB acquired an ability to produce a higher butyrate concentration that could not be achieved by the original wild-type culture grown in suspension.

Butyrate Tolerance of Mutants

Butyric acid is inhibitory to cell growth. To determine if there were any phenotypic changes about butyrate tolerance in mutant strains, cells were grown as suspension cultures at various initial butyrate concentrations (0~15 g/L). The specific growth rates were measured and shown as the relative growth rate compared to those obtained without added butyrate in the medium.

As shown in Figure 3, *ack* and *pta* deleted mutants had higher butyrate tolerance than the wild-type. At 15 g/L of butyric acid, both mutants retained ~ 30% of their maximum growth rates but less than 10% in the wild-type. The enhanced butyrate tolerance in these mutants might have contributed to their higher butyrate productivity and final butyrate concentration obtained in the fermentation. It should be noted that PTA in *C. tyrobutyricum* was more strongly inhibited by butyric acid than PTB (data not shown). It is thus possible that by disrupting the butyrate-sensitive PTA and acetate-forming pathway, the mutants became less sensitive to butyrate inhibition since they only used the butyrate-forming pathway to generate ATP needed for biosynthesis and maintaining a functional pH gradient across the cell membrane.

For the overexpression mutant WT(pTHBUT), it was more sensitive to butyrate inhibition than the wild-type. Less than 9% of the maximum growth rate was retained when butyrate was only 5 g/L and no cell growth was observed in the medium when butyrate was higher than 15 g/L. This result explains the fact observed in the fed-batch fermentation that this mutant stopped butyrate production very early when butyrate concentration only reached ~12 g/L. This overexpressed butyrate operon originated from the solventogenic *C. acetobutylicum*. It was reported that butyrate was more toxic than butanol (28), and even butyrate-producing enzyme PTB was inhibited by butyric acid in *C. tyrobutyricum*. Since *C. acetobutylicum* was able to shift from acidogenesis to solventogenesis and reassimilate butyrate under appropriate conditions to alleviate the butyrate inhibitory effects (29), it is suggested that the enzymes involved in butyrate production from this bacterium may be less tolerant to

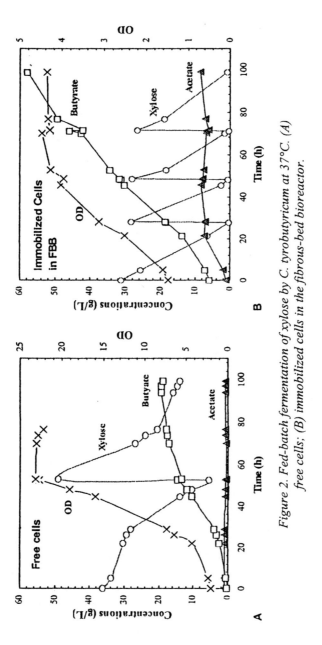

Figure 2. Fed-batch fermentation of xylose by C. tyrobutyricum at 37°C. (A) free cells; (B) immobilized cells in the fibrous-bed bioreactor.

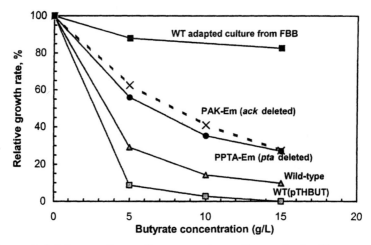

Figure 3. Inhibition effects of butyric acid on cell growth of wild-type and various mutant strains.

butyrate than those native ones from *C. tyrobutyricum*. Therefore, the exogenous butyrate operon on pTHBUT might be responsible for the reduced butyrate tolerance in WT(pTHBUT). Another reason for the reduced butyrate tolerance in WT(pTHBUT) might be the host-plasmid interactions (14). It was reported that the plasmid-bearing strains of *C. acetobutylicum* had slower growth rate, elevated solvent levels and lowered acid levels, possibly due to the induction of stress proteins in response to plasmid-imposed metabolic stress (14, 23). A similar stress response might have also occurred in WT(pTHBUT) and resulted in the decreased butyrate tolerance of cells.

It is also noticed that the adapted culture from the FBB was much less sensitive to butyrate concentration increase as compared with the wild-type culture, indicating a much higher tolerance to butyrate (Figure 3). This explains the higher butyrate concentration level obtained in the immobilized cell fermentation (Figure 2). The adapted cells retained more than 50% of its growth ability when the butyrate concentration increased to 30 g/L, whereas the original culture had lost its ability to grow at a butyrate concentration beyond 20 g/L (data not shown). Therefore, the ability to produce higher butyrate concentrations in the FBB can be attributed to the emergence of butyrate-tolerant mutant in the bioreactor through adaptation and natural selection, which did not happen in the suspension culture. Clearly, cells grown in fibrous matrices are more robust and have acquired the ability to tolerate higher butyrate concentrations. The adapted culture obtained should have a good potential in industrial butyrate production. It is expected that, by immobilization of genetically engineered mutants in the fibrous bed, their butyrate tolerance and production ability will be further improved and it will ultimately improve the process economics.

Conclusion

This study demonstrates that both genetic engineering and culture adaptation in the FBB can be used to develop mutant strains of *C. tyrobutyricum* and improve the process economics for butyrate fermentation. The adapted culture from the FBB is physiologically different from the original culture used to seed the bioreactor. This important finding has never been reported in conventional fermentation systems. The high productivity and high product concentration obtained in the FBB are either comparable or better than those reported in the literature. The manipulation of acid-forming pathways by gene inactivation and overexpression proved to be feasible for obtaining metabolically advantageous mutants for butyrate production from sugars. The fermentation kinetic studies of these mutants also provided valuable information about gene function in cellular metabolism, which can guide future effort to engineer novel super-producing strains of *C. tyrobutyricum* for industrial applications. The increased productivity and selectivity by mutant strains of *C. tyrobutyricum* immobilized in the FBB should lower the cost of bio-based butyrate and allow it to compete favorably in the market.

Acknowledgements

This work was supported in part by research grants from the Department of Energy-STTR (DE-FG02-00ER86106) and the U.S. Department of Agriculture (CSREES 99-35504-7800).

Reference

1. Sharpell, F. H. J. In *Comprehensive Biotechnology*, Moo-Young, M., Ed.; Pergamon Press: New York, 1985; pp 965-981.
2. Pouillart, P. R. *Life Sci*, **1998**, *63*, 1739-1760.
3. Huang, Y. L.; Wu, Z.; Zhang, L.; Cheuang, C. M.; Yang, S.-T. *Bioresource Technol.* **2002**, *82*, 51-59.
4. Michel-Savin, D.; Marchal, R.; Vandecasteele, J. P. *Appl Microbiol Biotechnol.* **1990**, *32*, 387-392
5. Zigova, J.; Sturdik, E.; Vandak, D.; Schlosser, S. *Process Biochem.* **1999**, *34*, 835-843.

6. McDaniel, R.; Licari, P.; Khosla, C. *Adv. Biochem. Eng. /Biotechnol.* **2001**, *73*, 31-52.
7. Aristidou, A.; Penttilä, M. *Curr Opin Biotechnol.* **2000**, *11*, 187-198.
8. Huang, Y.; Yang, S.-T. *Biotechnol. Bioeng.* **1998**, *60*, 499-507.
9. Huang, Y. L.; Mann, K.; Novak, J. M.; Yang, S. T. *Biotechnol. Progr.* **1998**, *14*, 800-806.
10. Silva, E. M.; Yang, S.-T. *J. Biotechnol.* **1995**, *41*, 59-70.
11. Yang, S.-T. U.S. Patent 5,563,069, 1996.
12. Yang, S.-T.; Zhu, H.; Li, Y.; Hong, G. *Biotechnol. Bioeng.* **1994**, *43*, 1124-1130.
13. Guérout-Fleury, A.-M.; Shazand, K.; Frandsen, N.; Stragier, P. *Gene* **1995**, *167*, 335-336.
14. Green, E. M.; Bennett, G. N. *Biotechnol. Bioeng.* **1998**, *58*, 215-221.
15. Rose, I. A. *Methods Enzymol.* **1955**, *1*, 591-595.
16. Andersch, W.; Bahl, H.; Gottschalk, G. *Eur. J. Appl. Microbiol. Biotechnol.* **1983**, *17*, 327-332.
17. Perego, M. In *Bacillus subtilis and other Gram-positive bacteria: biochemistry, physiology and molecular genetics*, Sonenshen, A. L.; Hoch, J. A.; Losick, R., Eds.; American Society for Microbiology: Washington D.C, 1993, pp 615-624.
18. Harris, L. M.; Desai R. P.; Welker, N, E.; Papoutsakis, E. T. *Biotechnol. Bioeng.* **2000**, *67*, 1-11.
19. Green, E. M.; Bennett, G. N. *Appl. Biochem. Biotechnol.* **1996**, *57-58*, 213-221.
20. Green, E. M.; Boynton, Z. L.; Harris, L. M.; Rudolph, F. B.; Papoutsakis, E. T.; Bennett, G. N. *Microbiol.* **1996**, *142*, 2079-2086.
21. Campbell, A. M. *Adv. Genet.* **1962**, *11*, 101-146.
22. Boynton, Z. L.; Bennett, G. N.; Rudolph, F. B. *Appl. Environ. Microbiol.* **1996**, *62*, 2758-2766.
23. Walter, K. A.; Mermelstein, L. D.; Papoutsakis, E. T. *Ann. N. Y. Acad. Sci.*, **1994**, *721*, 69-72.
24. Mermelstein, L. D.; Welker, N. E.; Bennett, G. N.; Papoutsakis, E. T. *Bio/Technology*, **1992**, *10*, 190-195.
25. Fayolle, F.; Marchal, R.; Ballerini, D. *J Ind Microbiol.* **1990**, *6*, 179-183.
26. Michel-Savin, D.; Marchal, R.; Vandecasteele, J. P. *Appl. Microbiol. Biotechnol.* **1990**, *34*, 172-177.
27. Zhu, Y.; Wu, Z.; Yang, S.-T. *Process Biochem.* **2002**, *38*, 657-666
28. Linden, J. C.; Moreira, A. *Basic Life Sci.* **1983**, *25*, 377-403.
29. Rogers, P.; Gottschalk, G. In *The clostridia and biotechnology*; Woods, D. R., Ed.; Butterworth-Heinemann: Stoneham, MA, 1993; pp 25-50.

Chapter 5

Production of Mannitol by Fermentation

Badal C. Saha

Fermentation Biotechnology Research Unit, National Center for Agricultural Utilization Research, Agricultural Research Service, U.S. Department of Agriculture, Peoria, IL 61604

Mannitol, a naturally occurring polyol or sugar alcohol, is widely used in the food, pharmaceutical, medicine, and chemical industries. The production of mannitol by fermentation has become attractive because of the problems associated with its production chemically. A number of heterofermentative lactic acid bacteria, yeasts, and filamentous fungi are known to produce mannitol. In this article, research dealing with mannitol production by fermentation using lactic acid bacteria, yeast, and fungi is presented. Several heterofermentative lactic acid bacteria are excellent producers of mannitol using fructose as an electron acceptor. Recent progress in the production of mannitol by fermentation and using enzyme technology as well as downstream processing of mannitol are described. The problems and prospects for mannitol production by fermentation and enzymatic means and future directions of research are highlighted.

Mannitol is a naturally occurring sugar alcohol found in many fruits and vegetables. It is found in pumpkins, mushrooms, onions, and in marine algae, especially brown seaweed. Brown algae contains about 10-20% mannitol depending on the harvesting time (*1*). Mannitol is present in fresh mushrooms at about 1%. It is widely used in the food, pharmaceutical, medicine, and chemical industries (*2*). Mannitol is used as a sweet-tasting bodying and texturing agent. It reduces the crystallization tendency of sugars and is used as such to increase the shelf-life of foodstuffs. Crystalline mannitol exhibits a very low hygroscopicity, making it useful in products that are stable at high humidity. It is only about half as sweet as sucrose. Mannitol exhibits reduced physiological calorie value (1.6 kcal/g) compared to sucrose (4 kcal/g). It has a low solubility in water of only 18% (w/v) at 25°C and 13% (w/v) at 14°C (*3*). In comparison, the solubility limit of sorbitol in water is about 70% (w/v) at 25°C. Mannitol is sparingly soluble in organic solvents such as ethanol and practically insoluble in ether, ketones, and hydrocarbons (*1*). It forms orthorhombic crystals and the crystals have melting point at 165-168°C (*1*). Mannitol is extensively used in chewing gum. It is chemically inert and is commonly used in the pharmaceutical formulation of chewable tablets and granulated powders. Mannitol prevents moisture absorption from the air, exhibits excellent mechanical compressing properties, does not interact with the active components, and has a sweet cool taste owing to its high negative heat of solution (approximately 121 kJ/kg) that masks the unpleasant taste of many drugs (*4*). The complex of boric acid with mannitol is used in the production of dry electrolytic capacitors. It is an extensively used polyol for production of resins and surfactants (*2*). Mannitol is used in medicine as a powerful osmotic diuretic (to increase the formation of urine in order to prevent and treat acute renal failure and also in the removal of toxic substances from the body) and in many types of surgery for the prevention of kidney failure (to alter the osmolarity of the glomerular filtrate) and to reduce dye and brain oedema (increased brain water content). Mannitol hexanitrate is a well known vasodilator, used in the treatment of hypertension (*5*).

Chemical Process for Production of Mannitol

Manna, obtained by heating the bark of tree *Fraxinus ornus*, can contain up to 50% mannitol and was the commercial source of mannitol for many years until the 1920s (*6*). Mannitol ($3.32 per pound) is currently produced industrially by high pressure hydrogenation of fructose/glucose mixtures in aqueous solution at high temperature (120-160°C) with Raney nickel as catalyst and hydrogen gas (Figure 1) (*6*). The β-fructose is converted to mannitol and the α-fructose is converted to sorbitol. The glucose is hydrogenated exclusively to sorbitol.

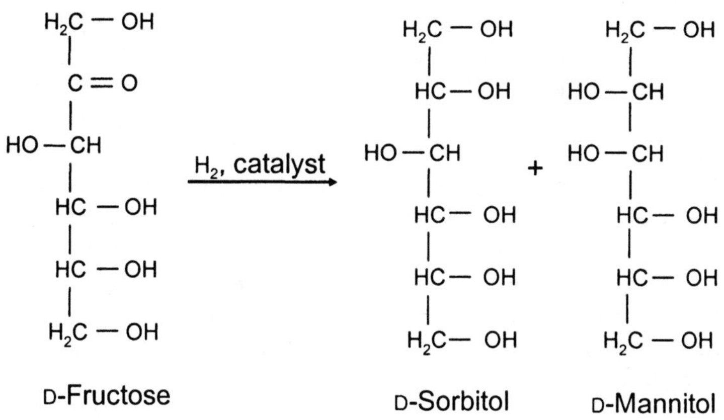

Figure 1. Catalytic hydrogenation of fructose.

Typically, the hydrogenation of a 50/50 fructose/glucose mixture results in an approximately 25/75 mixture of mannitol and sorbitol ($ 0.73 per pound). This is due to the poor selectivity of the nickel catalyst used. As a consequence, the commercial production of mannitol is always accompanied by the production of sorbitol, thus resulting in a less efficient process (2). Mannitol is less soluble than sorbitol and is generally recovered by cooling crystallization. According to Takemura et al. (7), the yield of crystalline mannitol in the chemical process is only 17% (w/w) based on the initial sugar substrates. If sucrose is used as starting material and the hydrogenation is performed at alkaline pH, mannitol yields up to 31% can be obtained (1). The hydrogenation of pure fructose results in mannitol yields of 48-50% (8).

Makkee et al. (9) developed a process involving both a bio- and a chemo-catalyst for the conversion of glucose/fructose mixture into mannitol. Good yields (62-66%) were obtained by using glucose isomerase immobilized on silica in combination with a copper-on-silica catalyst (water, pH ~7.0, 70°C, 50 kg/cm^2 of hydrogen, trace amounts of buffer, Mg (II), borate, and EDTA). In another method, mannitol is produced from mannose by hydrogenation with stoichiometric yield (100% conversion) (8). Mannose can be obtained from glucose by chemical epimirization with a yield of about 30-36% (w/w). Thus, the mannitol yield from initial sugar can be as high as 36%. Pure mannose is expensive. If the non-epimirized glucose can be enzymatically isomerized to fructose by using glucose isomerase, the mannitol yields could reach to 50% (w/w) (7). However, the total cost using the multi-steps process is not economical. Devos (8) suggested a process in which fructose is first isomerized to mannose using mannose isomerase. However, mannose isomerase is not yet commercially available for large scale use.

Microbial Production of Mannitol

In recent years, research efforts have been directed towards production of polyols by fermentation and enzymatic means (10). Lactic acid bacteria (LAB), yeasts and filamentous fungi are known to produce mannitol.

Bacteria

Several heterofermentative LAB belonging to the genera *Lactobacillus*, *Leuconostoc* and *Oenococcus* have been reported to produce mannitol effectively (11). In addition to mannitol, these bacteria may produce lactic acid, acetic acid, carbon dioxide and ethanol. Martinez et al. (12) reported that *Lb. brevis* fermented

1 mol of fructose to 0.67 mol of mannitol and 0.33 mol each of lactate and acetate. Soetaert et al. (*6,13*) reported a fed batch fermentation method with automatic feeding strategy for very fast and rapid production of mannitol and D-lactic acid from fructose or glucose/fructose mixture by using *Leu. pseudomesenteroides*. The maximal volumetric productivity of mannitol was 11.1 g/L.h with a final concentration of 150 g/L in 24 h and a conversion efficiency of 94%. Mannitol was also produced from fructose by *Leu. mesenteroides* immobilized in polyurethane foam with a yield of 1.0 and 8.0 g/L.h in batch and continuous fermentations, respectively (*14*).

Yun et al. (*15*) reported about 4-5 g of mannitol accumulation during fermentation of kimchi, a Korean pickled vegetable. Yun and Kim (*16*) isolated two different lactic acid bacteria, *Lactobacillus* sp. Y-107 and *Leuconostoc* sp. Y-002 during the fermentation of kimchi. These two strains utilized fructose and sucrose as substrates for mannitol formation. Under optimal conditions, the final mannitol concentration produced by *Lactobacillus* sp. Y-107 (at 35 °C, initial pH 8.0, anaerobic, 100 g fructose/L, 120 h) and *Leuconostoc* sp. Y-002 (at 35 °C, initial pH 6.0, anaerobic, 50 g fructose/L, 25 h) were 73 and 26 g/L from 100 g/L fructose with yields of 86 and 65% based on fructose consumed, respectively. The volumetric productivities of mannitol by both strains were less than 1.0 g/L.h. Neither isolate produced other polyols such as glycerol and sorbitol as by-products. These two bacterial strains were not able to use high concentrations of sugars above 100 g/L due to low osmotolerance of the isolates.

Erten (*17*) studied the utilization of fructose (5 mmol/L) as an electron acceptor in two *Leu. mesenteroides* strains under anaerobic conditions. These strains produced 0.26 mol mannitol, 0.65-0.67 mol of lactic acid, 0.37-0.57 mol ethanol and 0.26-0.27 mol acetic acid per mol of fructose at 25 °C. Fermentation of a mixture of fructose and glucose (1:1) resulted in the production of the same metabolic end-products. Korakli et al. (*18*) reported that mannitol is produced by sourdough *Lactobacilli* from fructose with concomitant formation of acetate. They obtained a 100% yield of mannitol from fructose by *Lb. sanfranciscensis* (isolated from sourdough) grown in a fed-batch culture containing fructose-glucose mixture with a volumetric productivity of 0.5 g/L.h and a final mannitol concentration of 60 g/L. The turnover of fructose was at its optimum at a concentration of glucose/fructose mixture ranging between 130-140 g/L in batch fermentation. Higher concentrations of substrate were found to be growth inhibitory for *Lb. sanfranciscensis*. After adaptation of the cells in sucrose, the bacterium produced mannitol to only 65% yield in relation to the fructose content of sucrose. It was shown that the bacterium synthesizes complex carbohydrates when grown on sucrose.

An unidentified *Lactobacillus* sp., named B001, produced mannitol from fructose with a volumetric productivity of 6.4 g/L.h (*19*). *Lb. pontis* isolated from sourdough produced mannitol, lactic acid, and ethanol from fructose (*20*). Salou et al. (*21*) reported that *O. oenos* converted 83 mol% of fructose to mannitol when grown in a medium containing fructose and glucose (1:1) with volumetric productivity of about 0.2 g/L.h. Most LAB are able to consume glucose and fructose simultaneously. Pimentel et al. (*22*) studied growth and metabolism of sugars and acids by *Leu. oenos* under different conditions of temperature and pH. The bacterium produced mannitol from fructose. The addition of acids, particularly, citrate, significantly repressed mannitol formation.

von Weymarn et al. (*11*) studied mannitol production by eight heterofermentative LAB - *Lb. brevis* ATCC-8287, *Lb. buchneri* TKK-1051, *Lb. fermentum* NRRL B-1932, *Lb. sanfranciscensis* E-93491, *Lactobacillus* sp. (B001) BP-3158, *Leu. mesenteroides* ATCC- 9135, *Leu. pseudomesenteroides* ATCC-12291, and *O. oeni* E-9762. They found that the ability to produce mannitol from fructose varied markedly among the heterofermentative LAB species. The effects of growth temperature, pH and nitrogen flushing on mannitol production by four selected strains were studied in batch bioreactor cultivation. Using *Lb. fermentum* and with fructose (20 g/L) and glucose (10 g/L) as carbon source, mannitol yields from fructose were 86.4±0.8, 88.9±2.4 and 93.6±0.6 mol% at 25, 30 and 35 °C, respectively. Mannitol yields but not the volumetric mannitol productivities were improved with constant nitrogen gas flushing of the growth medium. Applying the most promising strain (*Lb. fermentum*), high average and maximum mannitol productivities (7.6 and 16.0 g/L.h, respectively) were achieved. von Weymarn et al. (*23*) then compared the ability to produce mannitol from fructose by ten heterofermentative bacteria (eight from above, *Leu. mesenteroides* ATCC-8086 and ATCC 8293) in resting state. They achieved high mannitol productivity (26.2 g/L.h) and mannitol yield (97 mol%) in high cell density membrane cell-recycle culture using the best strain, *L. mesenteriodes* ATCC-9135. A stable high-level production of mannitol was maintained for 14 successive bioconversion batches using the same initial cell biomass. von Weymarn et al. (*23*) also reported that increasing the initial fructose concentration from 100 to 120 and 140 g/L resulted in decreased productivities due to both substrate and end-product inhibition of the key enzyme mannitol dehydrogenase (MDH). The by-products of this bioprocess were mainly acetate and lactate. Ojamo et al. (*24*) achieved a volumetric mannitol productivity of about 20 g/L.h using high cell density fermentation of *Leu. pseudomesenteroides* ATCC-12291.

Saha and Nakamura (*25*) found nine mannitol producing cultures after screening 72 bacterial cultures from the ARS Culture Collection on fructose. In addition, these cultures produced lactic acid and acetic acid. These strains are: *Lb brevis* NRRL B-1836, *Lb. buchneri* NRRL B-1860, *Lb. cellobiosus*

NRRL B-1840, *Lb. fermentum* NRRL B-1915, *Lb. intermedius* B-3693, *Leu. amelilibiosum* NRRL B-742, *Leu. citrovorum* NRRL B-1147, *Leu. mesenteroides subsp. dextranicum* NRRL B-1120 and *Leu. paramesenteroides* B-3471. The mannitol yields by *Lb. intermedius* B-3693 were 107.6 ± 0.5, 138.6 ± 6.9, 175.6 ± 5.9, and 198.3 ± 11.0 g/L at 150, 200, 250, and 300 g/L fructose, respectively in pH-controlled (pH 5.0) fermentation at 37°C. Small white needle-like crystals of mannitol appeared upon refrigeration of the cell-free fermentation broth of 300 g/L fructose at 4°C. The time of maximum mannitol yield varied greatly from 15 h at 150 g/L fructose to 136 h at 300 g/L fructose concentration. The bacterium transformed fructose to mannitol from the early growth stage. One-third of fructose can be replaced with other substrates such as glucose, maltose, starch plus glucoamylase (simultaneous saccharification and fermentation, SSF), mannose, and galactose. Two-thirds of fructose can also be replaced by sucrose. The bacterium co-utilized fructose and glucose simultaneously and produced very similar quantities of mannitol, lactic acid, and acetic acid in comparison with fructose. The glucose was converted to lactic acid and acetic acid, and fructose was converted into mannitol. Application of a fed-batch culture technique by feeding equal amounts of substrate and medium four times decreased the fermentation time from 136 h to 92 h to complete 300 g/L fructose utilization. The yields of mannitol, lactic acid, and acetic acid were 202.5 ± 4.3, 52.6 ± 0.96, and 38.5 ± 0.7 g/L, respectively. The bacterium utilized glucose (150 g/L) and produced D- and L-lactic acids in equal ratios (total, 70.4 ± 0.6 g/L) and ethanol (38.0 ± 0.9 g/L) but no acetic acid.

Several heterofermentative LAB produce mannitol in large amounts, using fructose as an electron acceptor. Mannitol produced by heterofermentative bacteria is derived from hexose phosphate pathway (*2,15,16,26*). The process makes use of the capability of the bacterium to utilize fructose as an alternative electron acceptor, thereby reducing it to mannitol with the enzyme MDH. In this process, the reducing equivalents are generated by conversion of one-third fructose to lactic acid and acetic acid (Figure 2). The enzyme reaction proceeds according to the following (theoretical) equation:

$$3 \text{ Fructose} \rightarrow 2 \text{ Mannitol} + \text{Lactic acid} + \text{Acetic acid} + CO_2$$

The net ATP gain is 2 mol of ATP per mol of fructose fermented.
For fructose and glucose (2:1) co-fermentation, the equation becomes

$$2 \text{ Fructose} + \text{Glucose} \rightarrow 2 \text{ Mannitol} + \text{Lactic acid} + \text{Acetic acid} + CO_2$$

For sucrose and fructose (1:1) co-fermentation, the equation becomes

$$\text{Sucrose} + \text{Fructose} \rightarrow 2 \text{ Fructose} + \text{Glucose} \rightarrow 2 \text{ Mannitol} + \text{Lactic acid} + \text{Acetic acid} + CO_2$$

Busse et al. (*27*) and Erten (*17*) found a lower mannitol yield from fructose in *Leu. mesenteroides*. In these cases, the enzyme reaction proceeds by the following equation:

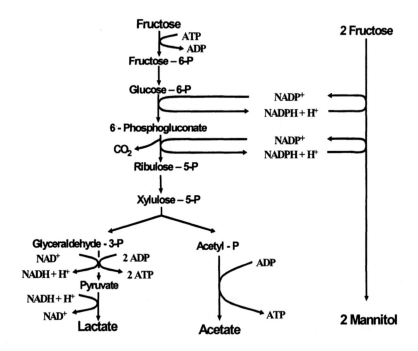

Figure 2. Pathway for production of mannitol from fructose by heterofermentative lactic acid bacteria.

(Reproduced from reference 25. Copyright 2003.)

3 Fructose → 1 Mannitol + 2 Lactic acid + 0.5 Acetic acid + 1.5 Ethanol + CO_2
The net gain is 1.25 mol of ATP per mol of fructose fermented..

Some homofermentative LAB such as *Streptococcus mutants* and *Lb. leichmanii* produce small amounts of mannitol *(28,29)*. Forain et al. *(30)* reported that a strain of *Lb. plantarum* deficient in both L- and D-lactate dehydrogenase (LDH) produces mannitol as an end-product of glucose catabolism. LAB uses several strategies for regeneration of NAD^+ during metabolism of carbohydrates. Veiga da Cumba et al. *(31)* reported that *O. oeni* produces erythritol to consume the reduced coenzymes formed in the glycolytic pathway. Hols et al. *(32)* showed that disruption of the *ldh* gene in *Lb. lactis* strain NZ20076 leads to the conversion of acetate into ethanol as a rescue pathway for NAD^+ regeneration. Neves et al. *(33)* reported that a LDH-deficient (LDH^d) mutant of *Lactococcus lactis* transiently accumulates intracellular mannitol, which was formed from fructose-6-phosphate (F-6-P) by the combined action of mannitol-1-phosphate dehydrogenase (MPDH) and phosphatase. They showed that the formation of mannitol-1-phosphate (M-1-P) by the LDH^d strain during glucose catabolism is a consequence of impairment in NADH oxidation caused by a highly reduced LDH activity, the transient formation of M-1-P serving as a regeneration pathway for NAD^+ regeneration. Grobben et al. *(34)* reported the spontaneous formation of a mannitol-producing variant of *Leu. pseudomesenteroides* grown in the presence of fructose. The mannitol producing variant differed from the mannitol-negative original strain in two physiological aspects: the presence of MDH activity and the simultaneous utilization of fructose and glucose. The presence of MDH is clearly a prerequisite for mannitol production. Kets et al. *(35)* found that the accumulation of mannitol is a typical feature of salt-stressed *Pseudomonas putida* strains grown in glucose mineral medium.

Yeasts

Some yeasts are known to produce a variety of sugar alcohols *(36)*. Onishi and Suzuki *(37,38)* studied the production of mannitol from glycerol by *Torulopsis* yeasts. *T. mannitofaciens* produced 31% mannitol from glycerol under optimal conditions. *T. versatilis* is also a good producer of mannitol from glycerol. Wako et al. *(39)* isolated a yeast strain (T-18) identified as *Torulopsis* sp. produced 23.7% mannitol and 42.4% glycerol in a medium (pH 9.0) containing 20% glucose. Hattori and Suzuki *(40)* studied the large-scale production of erythritol and its conversion of mannitol by *Candida zeylanoides* grown on alkane. The strain produced about 180 g/L meso-erythritol and a small amount of mannitol. Erythritol was almost entirely converted to mannitol by keeping the KH_2PO_4 concentration in the medium at 40-200 mg/L, and the yield was 63 g/L after 100 h incubation in a 5-L fermentor, which corresponded to 52% of the alkane consumed. *Zygosaccharomyces rouxii* ATCC 12572 produces ethanol, glycerol, arabitol, and mannitol from glucose (*41*).

Stankovic et al. (*42*) studied mannitol production from pentose sugars by *Rhodotorula minuta* (CCA 10-11-1). This was the only strain among 28 species of *Candida, Bretanomyces, Decceromyces, Kluyveromyces, Saccharomyces, Schwanniomyces* and *Trichosporon* that produced mannitol from D-pentose sugars. The yeast produced 16, 4, 5, and 5% mannitol from ribose, xylose, arabinose, and lyxose, respectively, when grown on these sugars (10%) at 28 °C and pH 4-7 for 14 days. In addition, the strain produced 3, 11, 5 and 6% D-arabinitol from these sugars, respectively. Song et al. (*43*) isolated over 1000 strains from various sources such as soil, seawater, honey, pollen and fermentation sludge and tested for their ability to produce mannitol. They identified a novel strain of *Candida magnoliae* (isolated from fermentation sludge) which produced 67 g/L mannitol in 168 h from 150 g/L fructose in batch flask culture (30 °C, 220 rpm, 10 g yeast extract/L). A fructose concentration higher than 200 g/L reduced the mannitol conversion yield and production rate. In fed-batch culture with 30-120 g fructose/L, mannitol production reached a maximum of 209 g/L after 200 h corresponding to 83% yield and a volumetric productivity of 1.03 g/L/h. The strain produced only small quantities of by-products, such as glycerol, erythritol and organic acids. De Zeeuw and Tynan (*44*) reported that *C. lipolytica* produces mannitol as main sugar alcohol. Yun and Song (*45*) reported that a strain of *Aureobasidium pullulans*, a yeast-like fungus, produced polyols of which mannitol was the main polyol associated with minute quantities of glycerol with a yield of about 23% (based on substrate utilized) from 20% (w/v) sucrose in batch flask culture at 30 °C and pH 6.0 in about 240 h. Stress solutes such as NaCl and KCl in the range from 0.25 to 1 M did not promote polyol production.

Filamentous Fungi

Several filamentous fungi produce mannitol from glucose. Yamada et al. (*46*) showed that glucose is first converted to F-6-P, which is then reduced to M-1-P in the presence of NADH, and M-1-P is hydrolyzed to mannitol by a specific phosphatase in *Pircularia oryzae*. Smiley et al. (*47*) studied the biosynthesis of mannitol from glucose by *Aspergillus candidus*. The fungal strain converted glucose to mannitol with 50% yield based on glucose consumed in 10-16 days by feeding glucose daily with a volumetric productivity of 0.15 g/L.h and a yield of 31.0 mol%. The presence of glucose in the medium was essential to prevent metabolism of mannitol. Nelson et al. (*48*) reported the production of mannitol from glucose and other sugars by conidia of *A. candidus*. Low pH (~ 3.0) favored the percentage yield but decreased the fermentation time. *A. Candidus* produced mannitol from 2% glucose in 75% yield (based on sugar consumed) in 7 days at 28 °C. The fungus converts glucose to mannitol via F-6-P and M-1-P (*49*). Lee (*50*) determined the carbon balance for fermentation of glucose to mannitol by *Aspergillus* sp. The products found were: cells (17% of carbon input), CO_2 (26%), mannitol (35%), glycerol (10%), erythritol (2.5%), glycogen (1%), and unidentified compounds (8%). Cell-free enzyme studies indicated that mannitol was produced via the reduction of F-6-P.

Hendriksen et al. (*51*) screened 11 different *Penicillum* species for production of mannitol. All strains produced mannitol and glycerol from sucrose. The highest amount of mannitol (43/L) was produced by *P. scabrosum* IBT JTER4 and the highest combined yield of mannitol and glycerol (65 g/L) was obtained with *P. aethiopicum* IBT MILA 4 when grown on sucrose (150 g/L) and yeast extract (20 g/L) at pH 6.2 and 25 °C for 12 days. However, the volumetric productivity of mannitol from sucrose by the high mannitol producer *P. scabrosum* was only 0.14/g/L.h. Foda et al. (*52*) investigated the conditions suitable for the maximum conversion of glucose to mannitol by *P. chrysogenum* Q 176 in submerged culture. The fungus produced 48.6% mannitol from glucose (50 g/L) after 6 days at 25-30°C and pH 7.0. *Penicillium* sp. uses the same metabolic route for conversion of glucose to mannitol as *A. candidus* (*53*). El-Kady et al. (*54*) screened 500 filamentous fungal isolates belonging to 10 genera and 74 species and identified *Aspergillus*, *Eurotium*, and *Fennellia* species as high (> 10 mmol/L) producers of mannitol cultivated on liquid glucose-Czapek's medium fortified with 15% NaCl and incubated at 28 °C as static cultures for 15 days. Domelsmith et al. (*55*) demonstrated that four fungal cultures - *Alternaria alternata*, *Cladosporium herbarum*, *Epicoccum purpurascens,* and *Fusarium pallidoroseum* isolated from cotton leaf dust produce mannitol and are a probable source of mannitol found in cotton dust. *A. niger* was reported to produce mannitol using sodium acetate as carbon source (*56*).

Enzymatic Production of Mannitol

Mannitol can be enzymatically produced from fructose in one pot synthesis by using NADH dependent MDH (EC 1.1.1.67) or NADPH dependent MDH (EC 1.1.1.38). The cofactor dependency of the enzyme is a major limitation. A number of strategies such as enzymatic, electrochemical, chemical and photochemical, and biological methods are available for cofactor regeneration (*57,58*). A two-enzyme system can be used for cofactor regeneration with simultaneous conversion of two substrates into two products of interest (*59*). One example is the simultaneous conversion of fructose and formate using the enzymes

MDH and formate dehydrogenase (FDH). FDH converts formate to CO_2 and reduces NAD to NADH. MDH uses NADH to convert fructose to mannitol and regenerates NAD.

Synthesizing reaction:

$$Fructose + NADH + H^+ \rightarrow Mannitol + NAD^+$$

Regenerating reaction:

$$Formic\ acid + NAD^+ \rightarrow CO_2 + NADH$$

Slatner et al. (60) achieved a volumetric mannitol productivity of 2.25 g/L.h using a recombinant MDH from *P. fluorescens* over-expressed in *Escherichia coli* and FDH from *C. boidinnii* with a final product concentration of 72 g/L and a fructose conversion of 80% in the system. The other product CO_2 is easily separable from mannitol. Mannitol was crystallized from the ultrafiltratered product solution in 97% purity and 85% recovery, thus allowing reuse of enzymes for repeated batch production of mannitol. Another example of the two enzyme system for regeneration of cofactor is MDH and glucose dehydrogenase (GDH) system using glucose-fructose mixture (1:1) and simultaneous synthesis of mannitol and gluconic acid (61). NAD requiring GDH converts glucose to gluconic acid and generates NADH. MDH uses NADH to convert fructose to mannitol and regenerates NAD.

Synthesizing reaction:

$$Fructose + NADH + H^+ \rightarrow Mannitol + NAD^+$$

Regenerating reaction:

$$Glucose + NAD^+ + H_2O \rightarrow Gluconic\ acid + NADH$$

The MDH used was from *P. fluorescens, Torulaspora delbruckii,* and *Schizophyllum commune,* and the FDH was from *Bacillus megaterium* (62,63). Downstream processing consists of the separation of enzymes from the product solution with the aid of reactor-integrated membranes and of the isolation of the two products-mannitol and gluconic acid by electrodialysis, ion-exchange chromatography, and fractional crystallization (64).

Both NADH- and NADPH-dependent MDHs have been purified from a number of microorganisms. As for example, NADH dependent MDH has been purified from *Lb. brevis, Leu. mesenteriodes, P. fluorescens, Rhodobacter spaeroides, Saccharomyces cerevisiae,* and *T. delbruckii* (13,65-69,63). NADPH dependent MDH has been purified from *A. parasitius, Zymomonas mobilis,* and *Gluconobacter suboxydans* (70-72). Schneider and Giffhorn (67) reported an increase of 8.3 fold of MDH activity by constructing a strain (pAK82) from *R. sphaeroides* Si4 and by producing high cell concentrations via fed-batch cultivation in a bioreactor in comparison to batch cultivation of the wild-type strain. Xylulose was produced by enzymatic conversion of D-arabitol using immobilized MDH from *R. sphaeroides* and methylene blue/O_2/diaphorase system for regeneration of NAD (73). The yield of the bioconversion was improved from 42 to 80% by preventing product inhibition of MDH complexing the produced xylulose with borate.

Mannose can be reduced to mannitol enzymatically. However, the reversible reaction favors mannitol oxidation and is thus not suitable (74).

Downstream Processing of Mannitol

von Weymarn et al. (23) crystallized mannitol with a yield of 72 mass % from permeates obtained from three successive batches of fermentation using *Leu. mesenteroides* ATCC-12291. After recrystallization, the purity of mannitol was over 99%. Deusing et al. (76) described a method for desalination by electrodialysis and crystallization of mannitol produced by large-scale fermentation by *Leu. mesenteroides* with a purity of > 99.5%. Soetaert et al. (77) reported that the use of electrodialysis followed by crystallization resulted in cost-effective recovery of highly pure crystalline mannitol and D-lactic acid under optimized downstream processing of the fermentation broth of *Leu. mesenteroides*. Slatner et al. (60) crystallized mannitol from an enzymatic conversion at 4°C (1 h) from the ultrafiltrate after evaporation of the product solution in *vacuo* to approximately half the original volume and subsequent addition of one volume equivalent of isopropanol. After centrifugation, the residual alcohol in the solid product was evaporated at room temperature, leaving mannitol with a purity of at least 97%. Ojamo et al. (24) suggested that the acidic side stream, obtained from a LAB fermentation process, could be used as feed preservative. Itoh et al. (20) used filtration to separate the cells from the fermentation broth. The cell-free broth was then concentrated by evaporation. Mannitol was first crystallized, and the crystals were separated by centrifugation. The mother liquor was further fractionated by chromatographic method into two fractions - acetate fraction and a lactate/mannitol fraction. Mannitol was separated from lactate by crystallization, and the lactate was isolated by precipitation with calcium hydroxide.

Concluding Remarks

Among the microorganisms reported to produce mannitol, heterofermentative LAB are promising candidates for production of mannitol by fermentation (Table I). These LAB have the capability to utilize high concentrations of fructose such that the mannitol concentration in the fermentation broth could reach more than 200 g/L, which is well enough to be separated as such from the cell-free fermentation broth by cooling crystallization. Most of the yeast and fungal cultures described in this chapter can utilize mannitol after it is made, which makes the

Table I. Mannitol Production by Fermentation

Organism	Substrate (g/L)	Time[a] (h)	Yield[b] (%)
Lactobacillus sp. B001(*19*)	Fructose (100) + Glucose (50)	24	65
Lactobacillus sp. Y-107 (*16,78*)	Fructose (100)	120	73
	Fructose (100)	75	71
Lb intermedius B-3693 (*25*)	Fructose (150)	20	68
Lb sanfranciscensis (*18*)	Fructose (?)	120	60 g/L
Leuconostoc mesenteroides (*2*)	Fructose (100) + Glucose (50)	35	60
Leuconostoc sp. Y-002 (*16*)	Fructose (50)	25	40
Leu. mesenteroides (*17*)	Fructose (08)	-	30-40
Candida magnoliae (*43*)	Fructose (150)	168	45
C. zeylannoides (*40*)	n-Paraffin (100)	100	52
Torulopsis mannitofaciens (*37*)	Glycerol (100)	168	31
T. versalitis (*36*)	Glucose (194)	240	28
Asergillus candidus (*47*)	Glucose (32)	288	69
Penicillium scabrosum (*51*)	Sucrose (150)	288	40

[a] Time to reach maximum mannitol yield.
[b] Mannitol yields were calculated on the basis of initial sugars employed.
From Ref. *25*

Table II. Comparison of Mannitol Production by Fermentation versus Catalytic Hydrogenation

Fermentation	Catalytic Hydrogenation
Substrate: fructose : glucose (2:1)	Substrate: fructose : glucose (1:1)
Fructose conversion to mannitol ~ 100%	Fructose conversion to mannitol ~ 50%
Co-product: lactic acid and acetic acid (half of mannitol)	Co-product: sorbitol (three times of mannitol)
No hydrogen gas required	Pure hydrogen gas is required
Nitrogen source and other nutrients required	Nickel catalyst required
Electrodialysis for removing organic acids	Ion exchange to remove nickel ions
Low purity raw materials may be used	High purity raw materials required
Yield of crystalline mannitol from initial sugar ~ 52%	Yield of crystalline mannitol from initial sugar ~ 39%
Side products formed per kg mannitol crystals produced ~ 0.7 kg	Side products formed per kg mannitol crystals produced ~ 1.6 kg
GRAS process (Food grade *Lactobacillus* sp.)	

(Reproduced from reference 25. Copyright 2003.)

process difficult to control, whereas most of the LAB do not consume any mannitol they produce. Regarding downstream processing of fermentation broth, mannitol can be recovered from cell-free fermentation broth by cooling crystallization at concentration above 180 g/L. The lactic acid and acetic acid can be recovered by electrodialysis (79). The enzyme MDH responsible for catalyzing the conversion of fructose to mannitol requires NADPH (NADH) as cofactor. It is thus possible to develop an one-pot enzymatic process for production of mannitol from fructose if a cost-effective cofactor regeneration system can be developed. Microbes have the advantage of a "built-in" cofactor regeneration machinery. The heterofermentative LAB cells can be immobilized in a suitable support, and the immobilized cells can be used in a bioreactor to continuously produce mannitol from fructose. The production of acetic acid and D-lactic acid by some LAB can be blocked (inactivation of acetate kinase and D-LDH) by mutagenesis. In that case, the fermentation broth should contain only mannitol and L-lactic acid. Both are value added-products and thus the fermentative production of mannitol will become more attractive. A summary of the advantages and disadvantages of using catalytic hydrogenation vs. fermentation for production of mannitol is presented in Table II.

References

1. Schwarz, E. In *Ulman's Encyclopedia of Industrial Chemistry;* Elvers, B.; Hawkins, S.; Russey, W., Eds.; VCH, Weinheim, **1994**, Vol. A25, 5th Edition, pp 423-426.
2. Soetaert, W.; Buchholz, K.; Vandamme, E. J. *Agro Food Ind Hi-Tech* **1995**, *6*, 41-44.
3. Perry, F. R.; Green, D. W.; Maloney, J. O. *Perry's Chemical Engineers' Handbook*, McGraw- Hill, New York, **1997**, 7th Edition, pp 2-40.
4. Debord, B.; Lefebvre, C.; Guypt-Hermann, A. M.; Hubert, J.; Bouche, R.; Guyot, J. C. *Drug Dev Ind Pharm* **1987**, *13*, 1533-1546.
5. Johnson, J. C. *Specialized Sugars for the Food Industry*, Noyes Data Corp., Park Ridge, **1976**, pp 313-323.
6. Soetaert, W. *Meded. Fac. Landbouwwet., Rijksumiv. Gent.* **1990**, *55*, 1549-1552.
7. Takemura, M.; Iijima, M.; Tateno, Y.; Osada, Y.; Maruyama, H. U.S. Patent 4,083,881, **1978**.
8. Devos, F. U.S. Patent 5,466,795, **1995**.
9. Makkee, M.; Kieboom, A. P. G.; van Bekkum, H. *Starch/Starke* **1985**, *37*, 136-141.
10. Vandamme, E. J.; Soetaert, W. *FEMS Microbiol. Rev.* **1995**, *16*, 163-186.
11. von Weymarn, N.; Hujanen, M.; Leisola, M. *Process Biochem.* **2002**, *37*, 1207-1213.
12. Martinez, G.; Barker, H. A.; Horecker, B. A. *J. Biol. Chem.* **1963**, *238*, 1598-1603.
13. Soetaert, W.; Buchholz, K.; Vandamme, E. J. *C. R. Acad. Agric. Fr.* **1994**, *80*, 119-126.
14. Soetaert, W.; Vandamme, E. J. *NATO ASI Ser., Ser A.* **1991**, *207*, 249-250.
15. Yun, J. W.; Kang, S. C.; Song, S. K. *J. Ferment. Bioeng.* **1996**, *81*, 279-280.
16. Yun, J. W.; Kim, D. H. *J. Ferment. Bioeng.* **1998**, *85,* 203-208.
17. Erten, H. *Proc. Biochem.* **1998**, *33*, 735-739.
18. Korakli, M.; Schwarz, E.; Wolf, G.; Hammes, W. P. Adv. Food Sci. **2000**, *22*, 1-4.
19. Itoh, Y.; Tanaka, A.; Araya, H.; Ogasawara, K.; Inaba, H.; Sakamoto, Y.; Koga, J. European Patent 486024, **1992**.
20. Hammes, W. P.; Stolz, P.; Ganzle, M. *Adv. Food. Sci.* **1996**, *18*, 176-184.
21. Salou, P.; Divies, C.; Cardona, R. *Appl. Microbiol. Biotechnol.* **1994**, *60*, 1459-1466.

22. Pimentel, M. S.; Silva, M. H.; Cortes, I.; Faia, A. M. *J. Appl. Bacteriol.* **1994**, *76*, 42-48.
23. von Weymarn, N.; Kiviharju, K.; Leisola, M. *J. Ind. Microbiol. Biotechnol.* **2002**, *29*, 44-49.
24. Ojamo, H.; Koivkko, H.; Hekkila, H. Patent appl. WO 00/04181, **2000**.
25. Saha, B. C.; Nakamura, L. K. *Biotechnol. Bioeng.* **2003**, in press.
26. Wisselink, H. W.; Weusthuis, R. A.; Eggink, G.; Hugenholt, J.; Grobben, G. J. *Int. Dairy J.* **2002**, *12*, 151-161.
27. Busse, M.; Kindel, P. K.; Gibbs, M. *J. Biol. Chem.* **1961**, *236*, 2850-2853.
28. Loesche, W. J.; Kornman, K. S. *Arch. Oral. Biol.* **1976**, *21*, 551-553.
29. Chalfan, Y.; Levy, R.; Mateles, R. I. *Appl. Microbiol.* **1975**, *30*, 476.
30. Forain, T.; Schauck, A. N.; Delcour, J. *J. Bacteriol.* **1996**, *178*, 7311-7315.
31. Veiga da Cumba, M.; Firme, P.; San Romao, M. V.; Santos, H. *Appl. Environ. Microbiol.* **1992**, *58*, 2271-2279.
32. Hols, P.; Ramos, A.; Hugenholtz, J.; Delcour, J.; de Vos, W. M.; Santos, H.; Kleerebezem, M. *J. Bacteriol.* **1999**, *181*, 5521-5526.
33. Neves, A. R.; Ramos, A.; Shearman, C.; Gasson, M. J.; Almeida, J. S.; Santos, H. *Eur. J. Biochem.* **2000**, *267*, 3859-3868.
34. Grobben, G. J.; Peters, S. W. P. G.; Wisselink, H. W.; Weusthuis, R. A.; Hoefnagel, H. N.; Hugenholtz, J.; Eggink, G. *Appl. Environ. Microbiol.* **2001**, *67*, 2867-2870.
35. Kets, E. P. W.; Galinski, E. A.; de Wit, M.; de Bont, J. A. M.; Heipieper, H. J. *J. Bacteriol.* **1996**, *178*, 6665-6670.
36. Onishi, H.; Suzuki, T. *Appl. Microbiol.* **1968**, *16*, 1847-1852.
37. Onishi, H.; Suzuki, T. *Biotechnol. Bioeng.* **1970**, *12*, 913-920.
38. Onishi, H.; Suzuki, T. *Hakko Kogaku Zasshi* **1970**, *48*, 563-566.
39. Wako, K.; Kawaguchi, G.; Kubo, N.; Kasumi, T.; Hayashi, K.; Iino, K. *Hakko Kogaku Kaishi* **1988**, *66*, 209-215.
40. Hattori, K, Suzuki, T. *Agr. Biol. Chem.* **1974**, *38*,1203-1208.
41. Groleau, D.; Chevalier, P.; Tse HingYuen, T. L. S. *Biotechnol. Lett.* **1995**, *17*, 315-320.
42. Stankovic, L.; Bilik, V.; Matulova, M. *Folia Microbiol. (Prague)* **1989**, *34*, 511-514.
43. Song, K. H.; Lee, J. K.; Song, J. Y.; Hong, S. G.; Baek, H.; Kim, S. Y.; Hyun, H. H. *Biotechnol. Lett.* **2002**, *24*, 9-12.
44. De Zeeuw, J. R.; Tynan, E. J. III. U.S. Patent 3, 736, 229, **1973**.

45. Yun, J. W.; Song, S. K. *Biotechnol. Lett.* **1994**, *16*, 949-954.
46. Yamada, H.; Okamoto, K.; Kodama, K.; Noguchi, F.; Tanaka, S. *J. Biochem. (Tokyo)* **1961**, *49*, 404-410.
47. Smiley, K. L.; Cadmus, M. C.; Liepins, P. *Biotechnol. Bioeng.* **1967**, *9*, 365-374.
48. Nelson, G. E. N.; Johnson, D. E.; Ciegler, A. *Appl. Microbiol.* **1971**, *22*, 484-485.
49. Strandberg, G. W. *J. Bacteriol.* **1969**, *97*, 1305-1309.
50. Lee, W. H. *Appl. Microbiol.* **1967**, *15*, 1206-1210.
51. Hendriksen, H. V.; Mathiasen, T. E.; Alder-Nissen, J.; Frisvad, J. C.; Emborg, C. *J. Chem. Technol. Biotechnol.* **1988**, *43*, 223-228.
52. Foda, I. O.; Abdel–Akter, M.; El-Nawawy, A. S. *J. Microbiol. UAR* **1966**, *1*, 97-115.
53. Boosaeng, V.; Sullivan, P. A.; Shepherd, M. G. *Can. J. Microbiol.* **1976**, *22*, 808-816.
54. El-Kady, L. A.; Moubasher, M. H.; Mostafa, M. E. *Folia Microbiol. (Prague)* **1995**, *40*, 481-486.
55. Domelsmith, L. N.; Klich, M. A.; Goynes, W. R. *Appl. Environ. Microbiol.* **1988**, *54*, 1784-1790.
56. Barker, S. A.; Gomez-Sanchez, A.; Stacey, M. *J. Chem. Soc.* **1958**, 2583-2586.
57. Chenault, H. K., Whitesides, G. M. *Appl. Biochem. Biotechnol.* **1987**, *14*, 147-197.
58. O'Neill, H.; Woodward, J. In *Applied Biocatalysis in Specialty Chemical and Pharmaceuticals*; Saha, B. C.; Demirjian, D. C., Eds.; American Chemical Society, **2000**; pp 103-130.
59. Wichmann, R.; Wandrey, C.; Buckmann, A. F.; Kula, M. R. *Biotechnol. Bioeng.* **1981**, *23*, 2789-2802.
60. Slatner, M.; Nagl, G.; Haltrich, D.; Kulbe, K. D.; Nidetzky, B. *Biocat. Biotrans.* **1998**, *16*, 351-363..
61. Howaldt, M.; Gottlob, A.; Kulbe, K.; Chmiel, H. *Ann. N. Y. Acad. Sci.* **1988**, *542*, 400-405.
62. Haltrich, D.; Nidetzky, B.; Miemietz, G.; Gollhofer, D.; Sabine, L.; Stolz, P.; Kulbe, K. D. *Biocat. Biotrans.* **1996**, *14*, 31-45.
63. Nidetzky, B.; Haltrich, D.; Schmidt, K.; Schmidt, H.; Weber, A.; Kulbe, K. D. *Biocat. Biotrans.* **1996**, *14*, 46-53.
64. Kulbe, K. D.; Schwab, U.; Gudernatsch, W. *Ann. N. Y. Acad. Sci.* **1987**, *506*, 552-568.

65. Sakai, S.; Yamanaka, K. *Biochim. Biophys. Acta* **1968**, *151*, 684-686.
66. Brunker, P.; Altenbuchner, J.; Kulbe, K. D.; Mattes, R. *Biochim. Biophys. Acta* **1997**, *1351*, 157-167.
67. Schneider, K. H; Giffhom, F. *Eur. J. Biochem.* **1989**, *1184*, 15-19.
68. Schneider, K. H.; Giffhorn, F.; Kaplan, S. *J. Gen. Microbiol.* **1993**, *139*, 2475-2484.
69. Quain, D. E.; Boulton, C. A. *J. Gen. Microbiol.* **1987**, *133*, 1675-1684.
70. Niehaus, W. G. Jr.; Dilts, R. P. *J. Bacteriol.* **1982**, *1151*, 243-250.
71. Viikari, L.; Korhola, M. *Appl. Microb. Biotechnol.* **1986**, *24*, 471-476.
72. Adachi, O.; Toyama, H.; Matsushita, K. *Biosci. Biotechnol. Biochem.* **1999**, *63*, 402-407.
73. Schwartz, D.; Stein, M.; Schneider, K. H.; Giffhorn, F. *J. Biotechnol.* **1994**, *33*, 95-101.
74. Stoop, J. M. H.; Williamson, J. D.; Conkling, M. A.; MacKay, J. J.; Pharr, D. M. *Plant Sci.* **1998**, *131*, 43-51.
76. Deusing, I.; Buchholz, K.; Bliesener, K. *Chem.-Ing.-Tech.* **1996**, *68*, 1436-1438.
77. Soetaert, W.; Vanhooren, P. T.; Vandamme, E. J. *Methods Biotechnol.* **1999**, *10 (Carbohydrate Biotechnology Protocols)*, 261-275.
78. Yun, J. W.; Kang, S. C.; Song, S. K. *Biotechnol. Lett.* **1996**, *18*, 35-40.
79. Datta, R, Tsai, S-P. In *Fuels and Chemicals from Biomass*; Saha , B. C.; Woodward, J. Eds.; American Chemical Society: Washington, DC, **1997**; pp 224-236.

Production of Pharmaceuticals

Chapter 6

Fermentation Process Development of Carbohydrate-Based Therapeutics

Chin-Hang Shu

Department of Chemical and Materials Engineering, National Central University, Chung-Li, Taiwan 320

This chapter focuses on the latest advances in the development of fermentation processes in carbohydrate-based therapeutics. Although fermentation has been the major route in producing bioactive compounds historically, the complexity of carbohydrates and the current remarkable accomplishments of glycobiology have forced a multidiscipline approach to the development of carbohydrate-based therapeutics. Current remarkable achievements of glycotechnology have been demonstrated in metabolic engineering, process optimization, analytical chemistry, monitoring techniques, enzymatic/ chemical synthesis, and separation techniques, etc. By intentionally incorporating these techniques into fermentation processes, numerous novel approaches have been generated to produce carbohydrate-based therapeutics with higher qualities in terms of authenticity and efficacy, and better yields. Besides, many novel bioactive compounds or drug candidates can be obtained by fermentation in a relatively economic approach.

Carbohydrates are considered traditionally as the major food ingredients for energy source. Due to the diversity and complexity of carbohydrates, scientists hinder the realization of the role of carbohydrates in the medical science in the last few decades. Recently, the situation has changed and carbohydrate research is becoming an important research subject for medicinal and pharmaceutical chemistry due to the dramatic advances in the knowledge of glycoproteins, glycopeptides, oligosaccharides and polysaccharides. Significant progress in the science of glycobiology has revealed the important roles of carbohydrates as the recognition molecules on cell surfaces for protein targeting signal and as receptors for binding toxins, viruses and hormones (1-4).

Simultaneously, numerous excellent reviews and books regarding the chemical synthesis have been published elsewhere (5-7). Consequently, the noteworthy progress in the science of glycochemistry has driven the production of carbohydrates for medical applications by chemical synthesis.

Likewise, many reviews focusing on utilizing enzymes in carbohydrate synthesis are emerging (8-10). Since most of the complex bioactive carbohydrates are produced in nature by enzymes, specific enzymes capable of regioselective transfer and high glycosylation yields have been characterized and used for synthesizing carbohydrate-based therapeutics.

Fermentation has been known as the major route for production of numerous antibiotics. However, the information of the production of carbohydrate-based therapeutics by fermentation technology in literatures is relatively scattered and rare. Most efforts of the fermentation technology have been focused on producing better glycoproteins, rather than producing pure carbohydrates. The main objectives of this chapter are to give an overview of the current development of fermentation technology in carbohydrate-based therapeutics and to point out the challenges and future development. Within the limited goals of this chapter, the following sections are basically trying to answer the following questions:

1. What are the current carbohydrate-based therapeutics and how they are made?
2. Why do these carbohydrates play the important role in cell-cell signaling medicines?
3. What are the current engineering achievements and challenges in the fermentation processes of carbohydrate-based therapeutics?

Current Carbohydrate-based Therapeutics

A selected list of carbohydrate-based therapeutics is shown in Table 1. Considering the chemical classes, carbohydrate-based therapeutics might be classified into two categories: pure carbohydrates and glycoconjugates, consisting of carbohydrates pared with peptides, proteins or lipids. The pure carbohydrate therapeutics includes acarbose, voglibose, miglitol, zanamivir etc.

Table 1: Current and Potential Carbohydrate-based Therapeutics

Drug name	Chemistry Type	Description	Synthetic Methods	Ref.
Acarbose	Pseudo-oligosaccharides	α-glycosidase inhibitor; antidiabetics	F; C	11
Amphotericin B	Macrolide	Antifungals	F; C;	12
Calicheamicin	Aminoglycoside	Antitumor antibiotics	F	13
Cerezyme (imiglucerase)	Glycolipid	Treatment of Gaucher's disease	F	14
Doxorubicin	Anthracycline	Anticancer drug	FC	15
Erythromycin	Macrolide	Antibiotics	F	16
Haemophilus b conjugate	Glycoproteins	Vaccine for *Haemophilus influenzae* b	F; FC	17
Heparin	Sulfated polysaccharide	Blood anticoagulant	Ex;	18
Lentinan	Polysaccharide	Antitumor	F	6
Miglitol	derivative of deoxynojiramycin	α-glycosidase inhibitor; antidiabetics	FC; C	19
Relenza (Zanamivir)	Sialic acid analogue	Neuraminidase inhibitor; Treatment of influenza	C	20
Sonifilan (Schizophyllan)	Polysaccharide	Antitumor	F	6
Topiramate	Fructose sulfamate	Anti-convulsant agent	F; C	21
Vancocin (Vancomycin)	Glycopeptides	Antibiotics	F;	22
Voglibose	Oligosaccharides	α-glycosidase inhibitor; antidiabetics	FC; C	6

C: chemical synthesis; E: enzymatic synthesis; F: fermentation; FC: fermentation/chemical methods; Ex: extraction

The glycoconjugate therapeutics includes erythromycin, Hib conjugates, vancomycin, and cerezyme. The carbohydrate moiety of glycoconjugates can influence the activity, clearance rate, antigenicity, immunogenicity, and solubility (*23, 24*).

Production Methods of Carbohydrate-based Therapeutics

Many carbohydrate-based drugs as listed in Table 1 are glycoproteins. Since the carbohydrate moiety of the glycoproteins are vital for their biological functions, the current production of these therapeutic glycoproteins is limited to mammalian cell cultures in spite of the well established techniques of heterologous protein production in other host systems such as insect cells, bacteria and yeasts. Nevertheless, this might change in the near future as many research breakthroughs are in progress as discussed in the latter part of this chapter.

As one of the top-selling carbohydrate drugs in the world since 1935, heparin, the anticoagulant sugar chain administered to prevent blood clots, is prepared by extraction from the pig intestinal lining. Nevertheless, commercial heparins are heterogeneous and poorly characterized mix of glycosaminoglycans with molecular weights ranging from 3000 to 30,000 Da. Thus, their potency and unwanted side effects vary from manufacturer to manufacturer. Recently, low-molecular-weight heparins (LMWHs) with molecular weight less than 6000 Da have been developed to reduce the unwanted side effects (*25*). Basically, there are three major strategies in generation of LMWHs, including controlled chemical cleavage, enzymatic digestion, and size fractionation of heparin (*18, 26*). While evaluating the potency of the diverse LMWHs from different manufacturers by different production strategies have driven scientists in understanding their structure and biological activity relationships and in achieving the goal of maintaining the homogeneousity of products. Thus, Dr. Sasisekharan and his colleagues in Massachusetts Institute of Technology are attempting to decipher the sequence of heparin's active side and to guide efforts in synthesizing potent LMWHs for specific medical applications (*22, 27*).

Although the preparation of heparins is by extraction, the following three routes, chemical synthesis, enzymatic synthesis, and fermentation or their combinations, as listed in Fig. 1, make most of the carbohydrate-based therapeutics. In the chemical route, regioselectivity is the prominent issue in carbohydrate chemistry and the undesired reactions of the hydroxyl groups of the sugars should be properly protected during reaction and deprotected afterwards as exampled by a classical disaccharide synthesis in Fig. 2 (*8*). In spite of the impressive development of novel and efficient methods for the synthesis of complex carbohydrates, these expensive protection and deprotection sequences might still limit their industrial applications (*8, 10, 28*). This might be avoided by using another route: enzymatic synthesis.

Carbohydrate production *in vitro* by enzymatic synthesis has the advantages of absolute regio- and stero-control of the configuration at the anomeric center. Therefore, the efforts and time required for the synthesis of carbohydrates is significantly reduced. Numerous carbohydrate-based therapeutics by enzymatic methods are thus created. However, as the structures of the interested carbohydrates become either too complex or too expensive to manufacture, scientists will resort to the *in vivo* technique: fermentation. The main advantages and disadvantages of these methods are summarized in Table 2. As the related technology and knowledge accumulated, combined methods sometimes might be better approach in the perspective of production or applications.

Polysaccharide based therapeutics such as lentinan, schizophyllan and Hib polysaccharide are mainly produced by fermentation due to the limitation of current technology of chemical or enzymatic methods in synthesizing large size of carbohydrates automatically. Since Hib vaccine is a type of bacterial capsular polysaccharide, which is type II T-cell independent antigen and is poorly immunogenic in children under 2 years of age. However, young children are the most susceptible to bacterial infections. Thus, Hib polysaccharide-protein conjugate vaccine is developed to convert the immune response to a T-cell dependent response, which can induce immunological memory and is immunogenic in very young child (*17*).

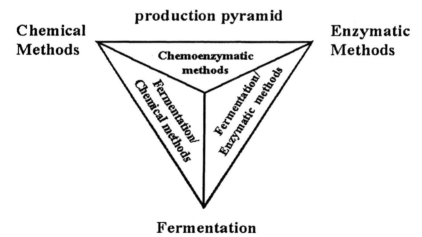

Fig. 1. Overview of the current production methods of carbohydrate-based therapeutics.

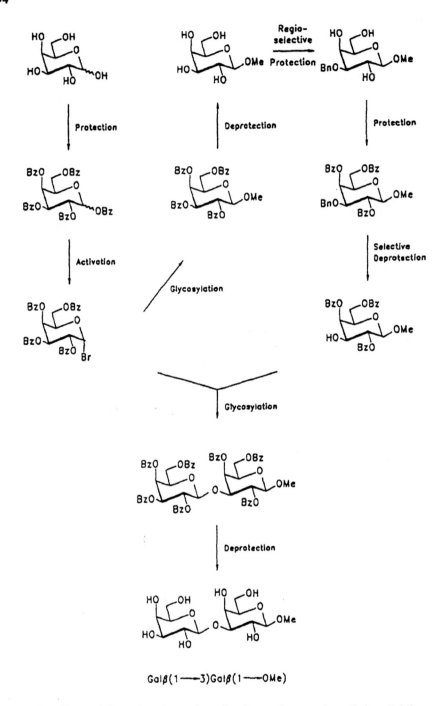

Galβ(1——3)Galβ(1——OMe)

Fig. 2. A classical disaccharide synthesis by chemical route: from Gal to Gal β, 1-3 Gal β1-OMe (8).

Table 2: Advantages and disadvantages of using varying approaches in production of carbohydrate-based therapeutics

Methods	Advantages	Disadvantages
Chemical synthesis	1. Capable of oligosaccharides synthesizers by developing an automated system 2. Capable of synthesizing complex and unnatural carbohydrates 3. Higher yields of synthesized carbohydrates	1. Extensive protection-deprotection schemes are required for the control of anomeric configurations. 2. Limitation of synthesizing larger size of carbohydrates
Enzymatic synthesis	1. Capable of catalyzing the synthetic reaction with stereo- and regioselectivity under very mild conditions. 2. Capable of producing more homogeneous products	1. High cost of enzymes and substrates 2. Lower yields of carbohydrates 3. Incapable of using unusual or unnatural sugars as substrates.
Fermentation	1. Capable of biosynthesis of complex carbohydrates 2. Sources of novel natural and bioactive compounds	1. Lower product yields 2. Costly downstream processing

Fundamentals of Carbohydrate Recognition Systems

Carbohydrate- protein interactions are the key of many important biological processes including signaling, recognition and catalysis. Lectins, carbohydrate-specific proteins, usually contain two or more carbohydrate-binding sites, which are known as carbohydrate recognition domains. The formation of carbohydrate-protein complexes usually initiates a cascade of further receptor-ligand interaction, leading to a biological event such as extravasations of leucocytes or signal transduction.

Lectins behave as receptors in diverse biological processes, including clearance of glycoproteins from the circulatory system, adhesion of pathogens to their host cells, recruitment of leucocytes to inflammatory sites, and in metastasis and malignancy. For example, in the cases of acute inflammation, the adhension of lecucocytes on endothelial layer of blood vessels and vascular exravasation, a process of the leucocyte migratin through the endothelial layer, have also been initiated by the interaction of carbohydrate and lectins on cell membranes as illustrated in Fig. 3 (*29*). Besides, the attachment of *Influenza* viruses to the glycosylated surface of their host cells is also lectin-mediated. The virus expresses two proteins on its surface to facilitate the virus to succeed in infecting a cell. One protein is a lectin, which is specific for sialic acid residues. It mediates adhesion of the virus to the host cell surface. The other protein is neuraminidase, which promotes the virus release from infected cells and facilitates virus spread within the respiratory tract. Zanamivir, the carbohydrate-based therapeutics, is a highly specific inhibitor of influenza neuraminidase activity, and prevent virus from replication (*20*).

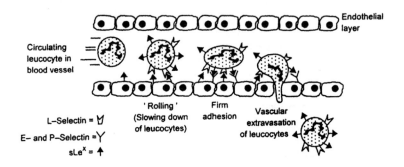

Fig. 3. A simplified representation of leucocyte adhension to endothelial cells and vascular exravasation in cases of inflammation (29).

(Reproduced with permission from reference 29. Copyright 2000 Wiley VCH)

Engineering Perspectives of Glycoconjugate Fermentation

There are many factors affecting the biosynthesis of carbohydrates of glycoconjugates such as the cell types, medium composition, and environmental conditions. Although comprehensive studies regarding the research area have not been reported, numerous significant findings of glycoproteins have been achieved as follows.

Choose of Appropriate Hosts and Media

Eukaryotic expression systems are required for complex recombinant proteins. Among the eukaryotic hosts, yeasts including *Saccharomyes* and *Pichia* can perform only mannosylated glycosylation, and plants do not have the enzymes to process N-linked glycosylation of oligosaccharide chains terminating with β-1, 4 galactose and /or sialic acid residues (*30, 31*). Thus, animal cell or incest cell cultures have opened the door for the availability of a multitude of therapeutic proteins that required complex post-modification for their biological functions.

Paradoxically, the authenticity of these recombinant proteins is still a major problem by using the biosynthetic route in the *in vitro* cell culture cultivation. Glycosylation is a common post-modification process influencing the authenticity of recombinant glycoproteins. In other words, natural glycoproteins usually exist as a spectrum of glycosylated forms (glycoforms), where each protein might associated with an array of oligosaccharide structures. There are several engineering approaches developed in improving or monitoring the glycosylation patterns of therapeutic glycoproteins.

Choice of complex or defined medium for recombinant protein proteins is based not only on yields of proteins but also on the quality of proteins and the cost of downstream processing. Serum was used as a growth-promoting component in the animal cell culture medium because it was a source of nutrients, hormones, growth factors and protease inhibitors. However, serum has several disadvantages. It is a mixture with respect to its chemical composition and many undesired components are present. For example, the bovine serum can be contaminated with viruses causing transmissible spongiform encephalopathy (TSE) and bovine spongiform encephalophathy (BSE) in human. Besides, high protein content of serum-containing culture media will result high cost and difficulty in downstream processes sing. Thus, adding meat hydrolysates has developed serum-free culture medium. However, these serum replacements still suffer from similar problems to those of serum (*32*). As a result, a chemically defined, protein-free, culture medium, has been developed to avoid the potential risk of introducing foreign agents or contamination derived from raw materials. The main challenge in designing a protein-free medium composition is to find the replacements of protein used for conventional culture media. In addition to the nutritional roles of proteins might play in the medium, the iron-carrier transferrin is replaced by EDTA or citrate (*33*), and the cell protection property of proteins from shear damage is replaced by the surfactant Pluronic F68 (*34*).

Medium Optimization

Besides the choice of medium, the chemical compositions have substantial influence on glycosylation and cell metabolism. For example, the glycosylation of the recombinant glycoprotein of NS0 and CHO cells might be manipulated by controlling their intracellular nucleotide-sugar contents through medium supplementation with the nucleotide sugar precursors glucosamine and N-acetylmannosamine (*35, 36*). Precursor studies indicate that the aromatic rings of the aglycone of macrolides are composed of two unusual amino acids that are metabolically derived from tyrosine and acetate, respectively (*37*). Thus, sufficient precursor supply by adding nitrogen sources such as peptone and amino acids into the medium might improve specific macrolide production such as vancomycin, but phosphate plays a negative control on macrolide and aminoglycoside antibiotics (*38- 40*).

Another fermentation strategy of medium optimization is to add inhibitors to block the biosynthetic pathways of undesired by-products. For instance, amphotericin B high producing strains of *Streptomyces nodosus* usually produce substantial more amphotericin A than amphotericin B as a result of costly downstream processing. A parallel biosynthetic pathway in the formation of amphotericin A and B is suggested upon biosynthetic studies, and a patented fermentation process is proposed to add chloramphenicol, a protein synthesis inhibitor, into the media to prevent the formation of amphotericin A, whereas amphotericin B production is slightly affected (*12*).

Process Optimization

Environmental factors including dissolved oxygen concentration, osmotic pressure, and carbon dioxide concentration have significant effects on the biosynthesis of oligosaccharides and glycoproteins. Some macrolide antibiotics including erythromycin B (*41*) and doxorubicin (*15, 42*) are considered to be secondary metabolites, and the limitation of the dissolved oxygen during cultivation of various microbial strains can decrease the activity of cytochrome P-450 monooxygenases required for the processing of pathway intermediates into their final forms. It is believed that the increase in DO concentration will affect the changes in energy metabolism with a higher proportion of glucose utilized at higher oxygen concentrations. However, some results from batch fermentation contradicted each other as shown below. Sialyltransferase activity, sialic acid content, and specific productivity of a recombinant glycoprotein, human follicle stimulating hormone, in CHO cells all increased as DO increased (*43*). In a well-controlled bioreactor, the glycan processing of the recombinant human secreted alkaline phosphatase in insect cell-baculovirus system was inhibited at both low and high dissolved oxygen concentrations (*44*). In contrast, another study indicated that the N-glycosylation of an interleukin-2 variant from BHK cells did not changed under low oxygen concentration (*45*). In a continuous culture study, alternations of the steady state DO concentration in the serum-free hybridoma culture dramatically affected the galactosylation of the monoclonal antibody (*46*). Their research results indicated that a significant decrease in galactosylation of at 10% DO as compared to those under 50 and 100%.

Carbon dioxide is a metabolic byproduct of mammalian cell culture that can accumulate in a poorly ventilated culture. High carbon dioxide partial pressure is likely to affect intracellular pH even though the medium pH is controlled. High intracellular pH might alter the endoplasmic reticulum and the Golgi apparatus. Changes in the pH of these organelles could alter the protein processing including glycosylation (47). Under elevated carbon dioxide concentration, the proportion of sialic acids comprising N-glycolylneuraminic acid of recombinant tissue plasminogen activator decreased from 2.3-4% under 36 mmHg CO_2 to 1.5-2.2% under 250 mmHg CO_2 (48).

Considering the production of oligosaccharides such as acarbose by fermentation, the process and product yields should be significantly optimized and improved in order to be economic competitive. The osmolarity of the fermentation broth has been found to have a very considerable effect on the final yield of the acarbose fermentation (11). Optimal range of osmolality for acarbose production is between 200 to 600 mosmol/kg. Under osmolarity-controlled fermentation, it was estimated to have 31% increases in final yield.

Challenge and Solutions to Maintaining Homogeneous Glycoforms of Glycoproteins

Heterogeneous glycoforms of recombinant proteins in mammalian cell cultures remains a common and complicated problem faced in current pharmaceutical. A simplied diagram of the synthesis of N-linked glycan is shown in Fig. 4 (49).

Heterogeneity of glycoforms of recombinant carbohydrate-based therapeutics has affected not only the efficacy of product but also the production cost. There are several engineering approaches developed in improving or monitoring the glycosylation patterns of therapeutic glycoproteins. Recombinant human erythropoietin (rhEPO), manufactured by Amgen, is discarded 80% for years because of its inadequate glycosylation, which causes rapid clearance from the blood (22). The pharmacokinetics of rhEPOs can be improved by adding two extra N-linked oligosaccharide chains to rhEPO to from a super-sialated erythropoietic protein, darbepoetin alpha (Aranesp). Aranesp has been proven with threefold greater circulating half-life than rhEPO (50, 51).

Incomplete synthesis of the N-linked carbohydrate structures on the recombinant glycoproteins is considered to the caused of heterogeneous glycoforms. Thus, the availability of sugars, the relative amounts of glycoprocessing enzymes, and the local protein structure around the glycosylation site in the cells will determine the glycoforms. Although several approaches including medium optimization, pathway engineering and process monitoring have been investigated to control the glycoforms, relatively little success in maintaining a homogeneous glycoform has been accomplished by using only fermentation techniques.

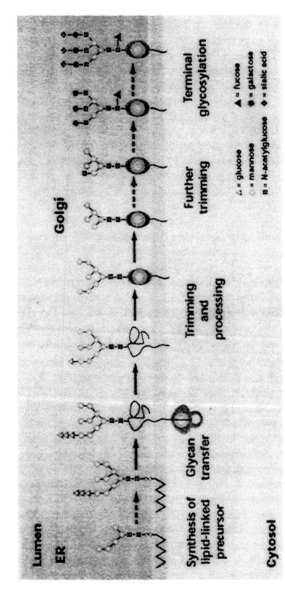

Fig. 4. A simplified diagram of the synthesis of N-linked glycan (49).

(Reproduced with permission from reference 49. Copyright 200 American Association of Sciences.)

A combined fermentation and enzymatic approach has been proposed to prepare the homogeneous glycoprotein as examplied by Fig. 5 (52, 53). Basically, the mixture of glycoproteins is used as the starting materials in which the sugar chain is digested enzymatically down to a simple homogeneous core and reelaborated enzymatically. In other words, this technology allows in vitro glycoprotein production by using enzymes to add sugars to protein in spite that its glycosylation is correct or not. It implies the possibility of using the starting materials from fungi with incorrect form of glycosylation. It also suggests a substantial slash of cost of recombinant glycoprotein production. Typically, the cost of producing a therapeutic glycoprotein from CHO cells will be between $1 million and $3 million per kilogram, and the cost of glycoprotein made in *Aspergillus* will be around $200~$300 per kilogram (53). Thus, it has drawn a great attention from the pharmaceutical and biotechnology companies. However, application of this approach is still underdeveloped and might be limited due to the size of proteins, the specific sequence of the oligosaccharides and the availability of the enzymes.

Another alternative of preparing the homogeneous glycoproteins is relying on appropriate separation methods. In the biological materials of culture broth such as cellular extracts and physiological fluids, glycoproteins are often present in minute quantities, placing high demands on the proper isolation procedures. A combination of general separation techniques and affinity principles are the most commonly practiced isolation/ fractionation strategies. Affinity chromatography has been developed into a powerful technique for purification of glycoproteins (54, 55). Affinity-based reverse micellar extraction and separation (ARMES) has also been developed to separate specific glycoproteins from mixtures of glycosylated (56). A highly purified rhEPO was obtained with a global yield of 50% as a result of combination techniques using an anti-EPO monoclonal antibody and DEAE-Sepharose (57). Since the microheterogeneity of rhEPO is a result of varying degree of sialylation of polysaccharidic chains, a pure glycoform can be obtained with the help of isoelectric focusing (IEF) (58).

The success of separation techniques in glycoproteins should also be accompanied by the significant progress of the analytical tools in complete elucidation of the microheterogeneous oligosaccharide structures of a glycoprotein. Usually, the analytical task is complex, time consuming and requiring multiple complementary techniques such as two-dimensional proton nuclear magnetic resonance, lectin affinity chromatography, and mass spectrometry (e.g., fast-atom bombardment or electrospray) or chromatography (e.g., gas chromatography or anion-exchange or size-exclusion liquid chromatography) performed with fragmentation techniques, such as methylation analysis, exoglycosidase treatment or, in the case of mass spectrometry, secondary ion analysis. Thus, there is a strong need in developing rapid analytical technique for monitoring the culture process. On-line harvesting and process monitor of recombinant proteins has been demonstrated (59). An analytical system for rapid assessment of site-specific microheterogeneity of the

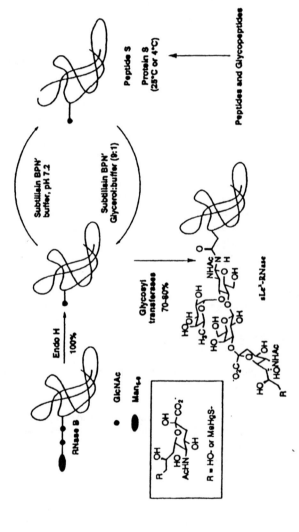

Fig. 5. A combined fermentation and enzymatic approach to the synthesis of homogeneous glycoprotein Ribonuclease B (52).

(Reproduced with permission from reference 52. Copyright 200 American Chemical Society.)

two potential N-linked glycosylation sites of recombinant human interferon- γ (IFN-γ) was proposed (*60*). The target protein is first purified by immunoaffinity chromatography, then isolated by an immobilized trypsin cartridge, then analyzed by matrix-assisted laser-desorption ionization/ time-of-flight mass spectrometry. By using this proposed analytical scheme, the analytical time is significantly reduced to 2 hours. On behalf of the achievement, monitoring the glycoforms is possible during the cultivational processes.

Metabolic Engineering

Metabolic engineering techniques are also utilized to modify the glycosylation pattern of glycoproteins in addition to monitoring the culture conditions for desired protein glycosylation patterns. The desired glycosylation patterns of glycoproteins can be achieved either by constructing a novel biosynthetic pathway through introducing genes encoding glycosyltransferases and glycosidases into the cells (*61- 63*) or by blocking a specific biosynthesis activity through using antisense technology (*64, 65*). Incorporation of two mammalian genes for α2, 6-sialyltransferase and β1, 4-galactosyltransferase into viral vectors has provided the lepidopteran insect cells with the new capability of sialylation of recombinant glycoproteins (*66*). Furthermore, overexpression of α2, 3-sialyltransferase and β1, 4-galactosyltransferase in CHO cells will result in maximizing sialic acid content of recombinant glycoproteins (*67*). Expression of antibodies with altered glycoforms by incorporating the gene for the expression of GnTIII in CHO cells leads to an increase in antibody-dependent cellular cytotoxicity through their higher affinity for FcγRIII (*68*). Besides providing the recombinants with the capabilities of high yields and authenticity of recombinant glycoproteins, metabolic engineering has opened a door for generating new drugs.

Several approaches such as combinatorial biosynthesis and engineering deoxysugar biosynthetic pathways have been developed for production of novel bioactive compounds. Although the natural products are the sources for the discovery of novel drugs, the possibility of finding new agents with desired antibiotic properties is getting lower. Thus, it is a relatively expedient and inexpensive approach in generation of novel glycosylated bioactive compounds by engineering the current biosynthetic pathways of antibiotics. For example, novel macrolides has been created by combinational biosynthesis, producing new unnatural or hybrid natural products (*69, 70*).

Since deoxysugars are the structural components of some important antibiotics (erythromycin), antifungal (amphotericin B), and anticancer drugs (doxorubicin), many bioactive compounds are created by constructing a plasmid (pLN2) which is able to direct the biosynthesis of different deoxysugars by exchanging and/ or adding genes from other antibiotic biosynthesis clusters (*16*). Simultaneously, the deoxysugars are transferred to the corresponding by using the elloramycin glycosyltransferase (ElmGT), which is known to be flexible in accepting different L- and D-deoxysugars.

Conclusion

Carbohydrates are becoming increasingly significant in medicine as knowledge of their pharmacology expands. In combination with the latest advance glycotechnology in enzymatic synthesis, chemical synthesis, analytical techniques, separation method, combinatorial biosynthesis and monitoring techniques, fermentation technology would make remarkable achievements in developing carbohydrate-based therapeutics as demonstrated partially in this chapter. The future of fermentation technology in carbohydrate-based therapeutics is promising especially in developing novel bioactive compounds.

References

1. Sharon, N.; Lis, H. *Chem. Eng. News* **1981**, *58*, 21.
2. Hakomori, S. I. *Spektrum. Wiss.* **1986**, *7*, 90.
3. Wiley, D. C.; Wilson, J. A.; Skeiiel, J. J. *Nature* **1981**, *289*, 373.
4. Varki, A. *Glycobiology* **1993**, *3*, 97-130.
5. *Glycoscience- synthesis of oligosaccharides and glycoconjugates*; Driguez, H.; Thiem, J., Eds.; Springer-Verlag, Berlin, Berlin, 1999.
6. *Carbohydrates in Drug Design*; Witczak, Z. J.; Nieforht, K. A., Eds.; Marcel Dekker, Inc., New York, 1997.
7. *Glycochemistry: Principles, Synthesis, and Applications*; Wang, P. G.; Bertozzi, C. R., Eds.; Marcel Dekker, Inc., New York, 2001.
8. Thiem, J. *FEMS Microbiol. Rev.* **1995**, *16*, 193-211.
9. Sears, P.; Wong, C. H. *Cell Mol. Life Sci.* **1998**, *54*, 223-252.
10. Scigelova, M.; Suddham, S., Crout, D. H. G. *J. Mole. Catalysis B: Enzymatic*, **1999**, *6*, 483-494.
11. Beunink, J.; Schedel, M.; Steiner, U. U.S. Patent 6,130,072, **2000**.
12. Schaffner, C. P.; Kientzler, D. J. U.S. Patent 6,132,993, **2000**.
13. Rothstein, D. M.; Love, S. F. *J. Bacteriol.* **1991**, *173*, 7716-7718.
14. Hoppe, H. *J. Biotechnol.* **2000**, *76*, 259-261.
15. Lomovskaya, N.; Otten, S.; Doi-katayama, Y.; Fonstein, L.; Liu, X.; Takatsu, T.; Inventi-Solari, A.; Filippini, S.; Torti, F.; Columbo, A; Hutchinson, C. R. *J. Bacteriol.* **1999**, *181*, 305-318.
16. Rodriguez, L.; Aguirrezabalaga, I.; Allende, N.; Brana, A. F.; Menez, C.; Salas, J. A. *Chem. Biol.* **2002**, *9*, 721-729.

17. Claesson, B. A.; Trollfors, B.; Lagergard, T.; Taranger, J.; Bryla, D.; Otterman, G.; Cramton, T.; Yang, Y.; Reimer, C. B.; Robbins, J. B. *J. Pediatr.* **1988**, *112*, 659-702.
18. Gunay, N. S.; Linhardt, R. J. *Planta. Med.* **1999**, *65*, 301-306.
19. Fouace, S.; Therisod, M. *Tetrahedon Lett.* **2000**, *41*, 7313-7315.
20. Gubareva, L. V.; Kaiser, L.; Hayden, F. *Lancet* **2000**, *355*, 827-835.
21. McComsey, D. F.; Maryanoff, B. E. *J. Org. Chem.* **1994**, *59*, 2652-2654.
22. Maeder, T. *Scientific Am.* **2002**, *7*, 40-47.
23. Goochee, C. F.; Gramer, M. J.; Andersen, D. C.: Bahr, J. B.; Rasmussen, J. R. In *Frontiers in bioprocessing II*; 1st ed.; Todd, P., Sikdar, S. F., Bier, M., Eds.; American Chemical Society; Washington, D. C., **1992**; pp 199-240.
24. Jenkins, N.; Curling, E. M. A. *Enzyme Microb. Technol.* **1994**, *16*, 354-364.
25. Weitz, J. I. *N. Engl. J. Med.* **1997**, 337, 688-698.
26. Casu, B.; Torri, G. *Semin. Thromb. Hemost.* **1999**, *25*, 17-25.
27. Shriver, Z.; Liu, D.; Sasisekharan, R. *Trends Cardiovasc. Med.* **2002**, *12*, 71-77.
28. Whitesides, G. M.; Wong, C. H. *Enzymes in Synthetic Organic Chemistry;* Elsevier, Oxford, **1994**, p. 252..
29. Lindhorst, T. K.. In *Essentials of carbohydrate chemistry and biochemistry*; Wiley-VCH, New York, **2000**; 185.
30. Montesino, R.; Garcia, R.; Quintero, O.; Cremata, J. A. *Protein Expression Purif.* **1998**, *14*, 197-207.
31. Rayon, C.; Lerouge, P.; Faye, L. *J. Exp. Bot.* **1998**, *49*, 1463-1472.
32. Merten, O. W. *Dev. Biol. Stand.* **1999**, *99*, 167-180.
33. Eto, N.; Yamada, K.; Shito, T.; Shirahata, S.; Murakami, H. *Agri. Biol. Chem.* **1991**, 55, 863-865.
34. Mizrahi, A. *Dev Biol. Stand.* **1984**, *55*, 93-102.
35. Baker, K. N.; Rendal, M. H.; Hills, A. E.; Hoare, M.; Freedman, R. B.; James, D. C. *Biotechnol. Bioeng.* **2001**, *73*, 188-202.
36. Hills, A. E.; Patel, A.; Boyd, P.; James, D. C. *Biotechnol. Bioeng.* **2001**, *75*, 239-251.
37. Lancini, G. C. *Prog. Ind. Microbiol.* **1989**, *27*, 283-296.
38. Fazeli, M. R.; Cove, J. H.; Baumberg, S. *FEMS Microbiol. Lett.* **1995**, *126*, 55-62.
39. McDermott, J.F.; Letherbridge, G.; Bushell, M. E. *Enzyme Microb. Technol.* **1993**, *15*, 657-668.
40. Mertz, F. P.; Doolin, L. E. *Can. J. Microbiol.* **1973**, *19*, 263-270.
41. Frykman, S. A.; Tsuruta, H.; Starks, C. M.; Regentin, R.; Carney, J. R.; Licari, P. *J. Biotechnol. Prog.* **2002**, 1-8.
42. Walczak, R.; Dickens, M.; Priestley, N.; Strohl, W. *J. Bacteriol.* **1999**, *181*, 298-304.

43. Chotigeat, W.; Watanapokasin, Y.; Mahler, S.; Gray, P. P. *Cytotechnol.* **1994**, *15*, 217-221.
44. Zhang, F.; Saarinen, M. A.; Itle, L. J.; Lang, S. C.; Murhammer, D. W.; Linhardt, R. L. *Biotechnol. Bioeng.* **2002**, *77*, 219-224.
45. Gawlitzek, M.; Conradt, H. S.; Wagner, R. *Biotechnol. Bioeng.* **1995**, *46*, 536-544.
46. Kunkel, J. P.; Jan, D. C. H.; Jamieson, J. C.; Butler, M. *J. Biotechnol.* **1998**, *62*, 55-71.
47. Thorens, B.; Vassalli, P. *Nature* **1986**, *321*, 618-620.
48. Kimura, R.; Miller, W. M. *Biotechnol. Prog.* **1997**, *13*, 311-317.
49. Helenius, A.; Abei, M. *Science* **2001**, *291*, 913-917.
50. Egrie, J. C.; Dwyer, E.; Lykos, M.; Hitz, A.; Browne, J. K. *Blood* **1997**, *90*, 56.
51. Elliott, S. G.; Lorenzini, T.; Strickland, T.; Delorme, E.; Egrie J. C. *Blood* **2000**, *96*, 82.
52. Witte, K.; Sears, P.; Martin, R.; Wong, C. H. *J. Am. Chem. Soc.* **1997**, *119*, 2114-2118.
53. Sears, P.; Wong, C. H. *Science* **2001**, *291*, 2344-2350.
54. Dakour, J.; Lundblad, A.; Zopf, D. *Anal. Biochem.* **1987**, *161*, 140-143.
55. Berglund, E.; Stigbrand, T.; Carlsson, S. R. *Eur. J. Biochem.* **1991**, *197*, 549-554.
56. Choe, J.; VanderNoot, V. A.; Linhardt, R. J.; Dordick, J. S. *AIChE J.* **1998**, *44*, 2542-2548.
57. Ben Ghanem, A.; Winchenne, J. J.; Lopez, C.; Chretien, S.; Dubarry, M.; Le Caer, J. P.; Casadevall, N.; Rouger, P.; Cartron, J. P.; Lambin, P. *Prep. Biochem.* **1994**, *24*, 127.
58. Gokana, A.; Winchenne, J. J.; Ben-Ghanem, A.; Ahaded, A.; Cartron, J. P.; Lambin, P. *J. Chromatogr. A* **1997**, *791*, 109-118.
59. Zhang, J.; Zhou, H.; Zhisong, J.; Regnier, F. *J. Chromatogr. B* **1998**, *707*, 257-265.
60. Harmon, B. J.; Gu, X.; Wang, D. I. C. *Anal. Chem.* **1997**, *68*, 1465-1473.
61. Grnbenhorst, E.; Hoffmann, A.; Nimtz, M.; Zettlmeissl, G.; Conradt, H. S. *Eur. J. Biochem.* **1995**, *232*, 718-725.
62. Grabenhorst, E.; Schlenke, P.; Pohl, S.; Nimtz, M.; Conradt, H. S. *Glycoconj. J.* **1999**, *16*, 81-97.
63. Monaco, L.; Marc. A.; Eon-Duval, A.; Acerbis, G.; Distefano, G.; Lamotte, D.; Engasser, J. E.; Soria, M.; Jenkins, N. *Cytotechnology* **1996**, *22*, 197-203.
64. Prati, E. G.; Scheidegger, P.; Sburlati, A. R.; Bailey, J. E. Biotechnol. Bioeng. **1998**, *59*, 445-450.
65. Prati, E. G.; Matasci, M.; Suter, T. B.; Dinter, A.; Sburlati, A. R.; Bailey, J. E. *Biotechnol Bioeng* **2000**, *68*, 239-244.

66. Jarvis, D. L.; Howe, D.; Aumiller, J. J. *J. Virol.* **2001**, *75*, 6223-6227.
67. Weikert, S.; Papac, D.; Briggs, J.; Cowfer, D.; Tom, S.; Gawlitzek, M.; Lofgren, J.; Mehta, S.; Chisholm, V.; Modi, N.; Eppler, S.; Carroll, K.; Chamow, S.; Peers, D.; Berman, P.; Krummen, L. *Nat. Biotechnol.* **1999**, *17*, 1116-1121.
68. Davies, J.; Jiang, L. Y.; Pan, L. Z.; Labarre, M. J.; Anderson, D.; Reff, M. *Biotechnol. Bioeng.* **2001**, *74*, 288-294.
69. Katz, L.; McDaniel, R. *Med. Res. Rev.* **1999**, *19*, 543-558.
70. Wohlert, S. E.; Blanco, G.; Lombo, F.; Fernandez, E.; Brana, A. F.; Reich, S.; Udvarnoki, G.; Mendez, C.; Decker, H.; Frevert, J.; Salas, J. A.; Rohr, J. *J. Am. Chem. Soc.* **1998**, *120*, 10596-10601.

Chapter 7

Submerged Fermentation of the Medicinal Mushroom *Ganoderma lucidum* for Production of Polysaccharide and Ganoderic Acid

Jian-Jiang Zhong, Ya-Jie Tang, and Qing-Hua Fang

State Key Laboratory of Bioreactor Engineering (China), East China University of Science and Technology, Shanghai 200237, China (email: jjzhong@ecust.edu.cn)

Effects of oxygen supply and lactose feeding on submerged cultures of *Ganoderma lucidum* were investigated. A higher $K_L a$ value led to a higher biomass density and a higher productivity of both intracellular polysaccharide and ganoderic acid. In a stirred bioreactor, at an initial $K_L a$ of 78.2 h^{-1}, a maximal cell concentration of 15.6 g/L by dry weight (DW) and a maximal intracellular polysaccharide (IPS) production of 2.2 g/L were obtained. An increase of initial $K_L a$ led to a higher production and productivity of ganoderic acid (GA), and the GA production and productivity at an initial $K_L a$ value of 96.0 h^{-1} was 1.8-fold those at an initial $K_L a$ value of 16.4 h^{-1}. In shake flasks, lactose feeding enhanced α-phosphoglucomutase activity, EPS production, phosphoglucose isomerase activity, and lactate accumulation. The information obtained may be useful for efficient large-scale production of these valuable bioactive products by the submerged cultures.

Mushrooms are abundant sources of a wide range of useful native products and new compounds with interesting biological activities. The obstacles in large-scale fermentation of mushroom include slow growth rate, low productivity, and difficulty in scale-up. In contrast to various studies on fermentation of conventional filamentous microorganisms such as streptomycetes and fungi, until now there are few investigations on the development of mushroom culture processes (*1*).

Ganoderma lucidum (Leyss.:Fr.) Karst is one of the most famous traditional Chinese medicinal mushrooms used as a health food and medicine in the Far East for more than 2000 years. Polysaccharides produced by *G. lucidum* are a type of carcinostatic agent, which have antitumor (*2*) and hypoglycemic activities (*3*). The higher fungus also produces ganoderic acids with various biological functions such as cytotoxicity to hepatoma cells, inhibition of cholesterol synthesis and stimulation of platelet aggregation (*1,4*), as well as new interesting biological activities including anti-tumor and anti-HIV-1 (*5,6*). Because it usually takes several months to cultivate the mushroom and the product yield is low in soil cultivation, submerged culture of *G. lucidum* is viewed as a promising alternative for efficient production of valuable polysaccharides (*7,8*) and ganoderic acids (*9,10*). However, there is scarce information on the simultaneous production of polysaccharide and ganoderic acid by submerged cultivation (*11-13*).

Oxygen (O_2) affects cell growth, cellular morphology, nutrient uptake, and metabolite biosynthesis. Ishmentskii et al. (*14*) reported that a high O_2 transfer rate could reduce, enhance, or have no effect on the production of pullulan, depending on the strain ploid of *Pullaria (Aureobasidium) pullulans*. In the submerged fermentation of *Monascus ruber,* Hajjaj et al. (*15*) reported that improving the O_2 supply increased the cell yield and production of red pigment and citrinin. Yoshida et al. (*16,17*) showed the significance of O_2 transfer in submerged cultures of a mushroom *Lentinus edodes*. Until now, there is lack of reports on effects of O_2 supply on simultaneous production of *Ganoderma* polysaccharide and ganoderic acid. One aim of this work was to study this factor.

In our preliminary work, lactose was found to be a favorable carbon source for the submerged fermentation of *G. lucidum*. Based on previous reports on EPS biosynthesis in bacteria (*18-20*), a possible pathway for our case is proposed in Fig.1. Enzymes leading to EPS formation can roughly be divided into four groups: enzymes responsible for the initial metabolism of a carbohydrate; enzymes involved in sugar nucleotide synthesis and interconversion; glycosyltransferases that form the repeating unit attached to the glycosyl carrier lipid; and translocases and polymerases that form the polymer. The phosphoglucose isomerase (PGI, Enzyme 8 in Fig.1) and α-phosphoglucomutase (α-PGM, Enzyme 3 in Fig.1) are the enzymes at the branch point between Embden-Meyerhof-Parnas (EMP) pathway and later part of EPS biosynthesis (Fig. 1). In another aspect, substrate feeding is a useful strategy to enhance cell density and process productivity. It is interesting and important to understand the

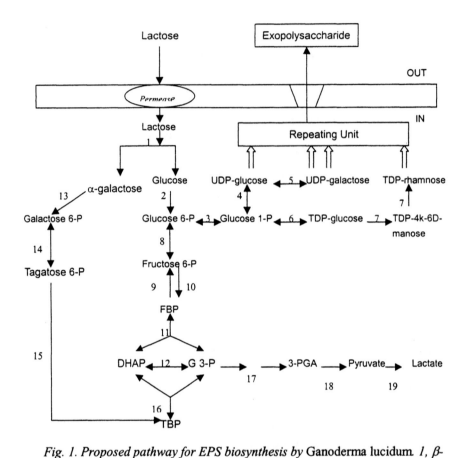

Fig. 1. Proposed pathway for EPS biosynthesis by Ganoderma lucidum. 1, β-galactosidase; 2, glucokinase; 3, α-phosphoglucomutase (α-PGM); 4, UDP-glucose pyrophosphorylase; 5, UDP-galactose-4-epimerase; 6, TDP-glucose pyrophosphorylase; 7, TDP-rhamnose biosynthetic system; 8, phophoglucose isomerase (PGI); 9, 6-phosphofructokinase; 10, fructose bisphosphatase; 11, fructose-1,6-bisphosphate aldolase; 12, triose phosphate isomerase; 13, α-galactose kinase; 14, galactose-6-phosphate isomerase; 15, tagatose-6-phosphate kinase; 16, tagatose-1,6-bisphosphate aldolase; 17, glyceraldehyde-3-phosphate dehydrogenase and phosphoglycerate kinase; 18, phosphoglyceromutase, enolase, and pyruvate kinase; and 19, lactate dehydrogenase. Abbreviations: DHAP, dihydroxyacetone phosphate; FBP, fructose 1,6-biphosphate; G 3-P, glyceraldehyde 3-phosphate; P, phosphate; 3-PGA, 3-phosphoglyceric acid; TBP, tagatose 1,6-bisphosphate; TDP, thymidine diphosphate; TDP-4K-6D-manose, TDP-4-keto-6-deoxymannose; UDP, uridine diphosphate.

responses of the product biosynthesis and related enzymes' activities to sugar feeding in *G. lucidum* fermentation. Another aim of this work was to investigate the effects of lactose feeding on the EPS production and the activities of related enzymes.

Materials and Methods

Maintenance and Preculture of *Ganoderma lucidum*

The strain *G. lucidum* was maintained on potato-agar-dextrose slants. The slant was inoculated and incubated at 28°C for 7 days, then stored at 4°C for about two weeks. Preculture medium and conditions were described elsewhere (*10-13*).

O_2 Supply Experiments

The effects of O_2 supply were preliminarily investigated by changing the medium loading volume of shake flask on a rotary shaker (30°C, 120 rpm). By setting the loading volume at 30, 50, 70 and 100 mL in 250-mL conical flasks, an initial K_La value of 32.6, 18.5, 13.2 and 6.4 h^{-1} was obtained, respectively. For bioreactor cultures, a 3.5-L (working volume) agitated bioreactor with two six-bladed turbine impellers (6.5 cm ID) was used. Cultivation was conducted at 30°C in the dark. The K_La value was determined using the dynamic gassing-in and gassing-out method (*21*). The cultures were all agitated at the same speed (200 rpm), and the aeration rate was set at 220, 1050, 1750 and 3500 mL/min to obtain the desired K_La values of 16.4, 60.0, 78.2 and 96.0 h^{-1}, respectively.

Lactose Feeding in Shake Flasks

The fermentation medium (with an initial pH of 5.5) was composed of 35 g L^{-1} of lactose, 5 g L^{-1} of peptone, 5 g L^{-1} of yeast extract, 1 g L^{-1} of KH_2PO_4, 0.5 g L^{-1} of $MgSO_4 \cdot 7H_2O$ and 0.05 g L^{-1} of vitamin B_1. The medium was inoculated with 5-mL second preculture broth (with ca. 350-400 mg DW of cells L^{-1}) in 45-mL medium in a 250-mL flask. The fermentation was conducted in the dark at 30°C on a rotary shaker at 120 rpm. Fed-batch fermentation was operated by feeding 15 g L^{-1} of lactose when its residual concentration was below 10 g L^{-1}. Samples were taken every 2-3 days. In all fermentations, there were 2-3 identical cultivation vessels operated under each condition, and the data shown represent average values with standard deviations.

Determination of Dry Weight, Residual Medium Sugar, and Lactate Concentration

For sampling, three flasks were sacrificed each time; for the bioreactors, about 20~30 mL of cell culture was taken once from each reactor. Dry cell weight (DW) and residual sugar concentration was measured by gravimetric method and phenol-sulfuric acid method, respectively (*11*). Lactate concentration in the fermentation broth was determined by using a conventional enzymatic method, and lactate dehydrogenase and NAD$^+$ were purchased from Sigma Chemical Co. (USA).

Measurements of *Ganoderma* Polysaccharide and Ganoderic Acid

The concentration of extracellular polysaccharide (EPS) in the fermentation broth, contents of intracellular polysaccharide (IPS) and ganoderic acid (GA) in mycelia were analyzed. Details of these analyses were reported earlier (*10-13*).

Preparation of Cell Extracts

Samples taken from fermentation broth were centrifuged (12,000×g, 10 min, 4°C) and the supernatant was decanted. The cell pellet was washed twice with 0.85% NaCl, and suspended in 20 mM phosphate buffer (pH 6.5) containing 50 mM NaCl, 10 mM MgCl$_2$, and 1 mM dithiothreitol. The cells were disrupted ultrasonically at 0°C for 72 cycles of 5s. Cell debris was removed by centrifugation (12,000×g, 10 min, 4°C). The protein concentration of the cell extract was determined by the method of Bradford (*22*) and compared with a bovine serum albumin standard.

Assay of β-Galactosidase, α-PGM and PGI Activities

β-Galactosidase (EC 3.2.1.23) activity was estimated by detecting the rate of hydrolysis of o-nitrophenyl-β-D-galactopyranoside (ONPG) at 30°C (*20*). The amount of ONPG hydrolyzed was determined by measuring the solution's absorbance at 420 nm. Assay of α-PGM (EC 2.7.5.1) and PGI (EC 5.3.1.9) was performed at 30°C in a total volume of 1.0 mL with freshly prepared cell extracts, and the formation or consumption of NAD(P)H was determined by measuring the absorbance change at 340 nm (*19*). The means of the results from at least two independent measurements were calculated.

Results and Discussion

Effects of O_2 Supply in Shake Flasks

Although there are a few studies on submerged cultures of *G. lucidum* for production of extracellular polysaccharides (*7,8*), no detailed information is available regarding the effects of O_2 supply on the accumulation of the fungus biomass and culture metabolites. As shown in Table 1, the maximum cell density of 14.6, 14.1, 11.9 and 11.5 g DW/L was obtained at day 6, 8, 10 and 12 at K_La of 32.6, 18.5, 13.2 and 6.4 h^{-1}, respectively. The results indicate that a higher O_2 supply led to more final biomass and shorter cultivation time. In fermentation of *Penicillium canescens* (*23*) and in cell culture of *Murine hybridomas* (*24*), it was also reported that good O_2 supply resulted in high cell growth rate.

For IPS formation, its highest content at an initial K_La of 32.6, 18.5, 13.2 and 6.4 h^{-1} was 6.73, 7.25, 7.93 and 8.06 mg/100 mg DW, respectively. The maximal production of IPS was obtained on day 6, 8, 10 and 14 at an initial K_La of 32.6, 18.5, 13.2 and 6.4 h^{-1}, respectively. Although the maximum IPS production was almost the same under the different O_2 supply conditions, a higher O_2 supply led to a higher IPS productivity. The maximal IPS production (and productivity) at an initial K_La of 32.6, 18.5, 13.2 and 6.4 h^{-1} was 0.83 (140), 0.87 (110), 0.86 (86) and 0.91 g/L (65 mg/(L d)), respectively (Table 1). The high IPS productivity under good O_2 supply was due to the high cell growth rate in the case. The EPS accumulation reached a peak value on day 8, 10, 10 and 12 at an initial K_La of 32.6, 18.5, 13.2 and 6.4 h^{-1}, respectively, and its maximal production titer was 0.52, 0.66, 0.63 and 0.64 g/L for each case. Their corresponding EPS productivity was 62, 63, 61 and 51 mg/(L d) (Table 1).

The highest content at an initial K_La of 32.6, 18.5, 13.2 and 6.4 h^{-1} was 1.39, 1.29, 1.54 and 1.45 mg/100 mg DW, respectively. The GA content of 1.54 mg/100 mg DW obtained here was more than 3-fold in a previous work by Tsujikura et al. (*9*). The maximum production of GA was obtained on the same day as the highest GA content. At an initial K_La of 32.6, 18.5, 13.2 and 6.4 h^{-1}, its highest production titer was 203, 180, 183 and 149 mg/L, respectively, and the corresponding GA productivity was 33.5, 22.2, 18.1 and 14.6 mg/(L d) for each (Table 1). A higher GA productivity was obtained under higher O_2 supply because of high cell growth rate in the case. The GA productivity of 33.5 mg/(L d) reached here was also much higher than that of previous work (*9*).

Table 1. Effects of loading volume on the cell growth, yield, and production of IPS, EPS and GA[a]

Loading volume (mL)	30	50	70	100
Cell dry weight (g/L)	14.6±0.0	14.1±0.2	11.9±0.6	11.5±0.2
Maximum IPS content (mg/100 mg DW)	6.7±0.2	7.3±0.1	7.9±0.1	8.1±0.1
Maximum IPS production (g/L)	0.83±0.02	0.87±0.07	0.86±0.07	0.91±0.06
IPS productivity (mg/(L d))	140	110	86	65
Maximum EPS production (g/L)	0.52±0.04	0.66±0.02	0.63±0.00	0.64±0.00
EPS productivity (mg/(L d))	62	63	61	51
Maximum GA content (mg/100 mg DW)	1.39±0.01	1.29±0.11	1.54±0.05	1.45±0.09
Maximum GA production (mg/L)	203±2	180±16	183±8	149±13
GA productivity (mg/(L d))	33.5	22.2	18.1	14.6

[a] Standard deviation was calculated from 3 samples

Effects of O_2 Supply in Bioreactors

Effect of initial K_La on *G. lucidum* cultures was also studied in bioreactors. All cell cultures were agitated at 200 rpm and aeration rate was adjusted over a range of 220 to 3500 mL/min to produce the desired initial K_La values from 16.4 to 96.0 h^{-1}. As shown in Table 2, initial K_La affected the biomass level, and its peak value of 15.6 g DW/L was obtained at an initial K_La of 78.2 h^{-1}. The results indicate that initial K_La had a significant effect on the cell growth during cultivation and an initial K_La of 78.2 h^{-1} seemed to be best for cell growth of *G. lucidum*. A typical time course of residual sugar concentration (at an initial K_La of 78.2 h^{-1}) is shown in Figure 2A.

The EPS production was 0.97, 0.69, 0.92 and 0.92 g/L in the cultures at initial K_La values of 16.4, 60.0, 78.2 and 96.0 h^{-1}, respectively, and its corresponding productivity was 73, 51, 69 and 69 mg/(L d) (Table 2). Figure 2B shows the kinetics of EPS accumulation at K_La of 78.2 h^{-1}. From day 0 to day 13, a rapid increase of EPS concentration was observed, and from day 13 to the end of culture (day 15), its accumulation level showed a slight decrease.

For the specific production (i.e., content) of IPS, as summarized in Table 2, the maximum IPS content of 17.5, 14.2, 14.0 and 15.6 mg per100 mg DW was reached on day 10 in the cultures at an initial K_La value of 16.4, 60.0, 78.2 and 96.0 h^{-1}, respectively. The maximum IPS production of 1.9, 1.6, 2.2 and 2.1 g/L was reached on day 10 in the culture grown at an initial K_La value of 16.4, 60.0, 78.2 and 96.0 h^{-1}, respectively, and its corresponding productivity was 0.19, 0.16, 0.22 and 0.21 g/(L d). The dynamic profile of the total accumulation of IPS is shown in Figure 2B. The results indicate that a relatively higher initial K_La value was favorable for IPS production and productivity.

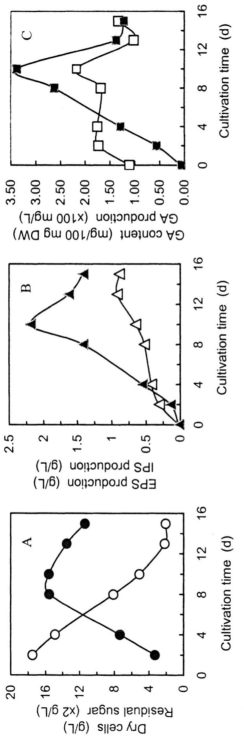

Figure 2. Time courses of dry cell weight (dark circle) (A), residual sugar concentration (open circle) (A), production of extracellular (EPS, open triangle) and intracellular polysaccharides (IPS, dark triangle) (B), and specific production (i.e., content, open square) and total production (dark square) of ganoderic acids (GA) (C) in submerged cultures of G. lucidum at initial K_{LA} of 78.2 h^{-1} in a bioreactor.

Table 2. Effects of initial O$_2$ transfer coefficient (K_La) on the cell growth, yield, and production of IPS, EPS and GA[a]

Initial K_La value (per h)	16.4	60.0	78.2	96.0
Cell dry weight (g/L)	11.8 (d 8)	14.1 (d 8)	15.6 (d 8)	12.9 (d 8), 13.6 (d 13)
Maximum IPS content (mg/100 mg DW)	17.5	14.2	14.0	15.6
Maximum IPS production (g/L)	1.9	1.6	2.2	2.1
IPS productivity (g/(L d))	0.19	0.16	0.22	0.21
Maximum EPS production (g/L)	0.97 (d 13)	0.69 (d 13)	0.92 (d 13)	0.92 (d 13)
EPS productivity (mg/(L d))	73 (d 13)	51 (d 13)	69 (d 13)	69 (d 13)
Maximum GA content (mg/100 mg DW)	2.33 (d 13)	2.44	2.17	3.36
Maximum GA production (mg/L)	245	280	338	450
GA productivity (mg/(L d))	23.8	27.3	33.1	44.3

[a] The maximal values were obtained on day 10 during cultivation except those indicated.

Figure 2C shows the kinetics of GA content and total GA production. Table 2 indicates that the maximum GA content in the culture at initial K_La values of 16.4, 60.0, 78.2 and 96.0 h^{-1} was 2.33, 2.44, 2.17 and 3.36 mg/100mg DW as obtained on day 13, 10, 10 and 10, respectively. Although the highest biomass was obtained at an initial K_La of 78.2 h^{-1}, the maximum GA production was obtained at an initial K_La of 96.0 h^{-1} because of the high GA content obtained in the latter case (Table 2). The total GA production of 245, 280, 338 and 450 mg/L was attained on day 10 in the culture grown at initial K_La values of 16.4, 60.0, 78.2 and 96.0 h^{-1}, respectively, and their corresponding productivity was 23.8, 27.3, 33.1 and 44.3 mg/(L·d). An increase in initial K_La led to an increased production and productivity of GA. The GA production and productivity at an initial K_La value of 96.0 h^{-1} was 1.8-fold those at an initial K_La value of 16.4 h^{-1}. It is clear that an initial K_La value of 96.0 h^{-1} was most suitable for both the production and productivity of GA.

Effect of Lactose Feeding on the Cell Growth and β-Galactosidase Activity

Figs. 3A and 3B show the cell growth (A) and residual sugar concentration (B) in batch and fed-batch fermentations of *G. lucidum* in shake flasks. The cell growth pattern was similar in both the flask and bioreactor fermentations. During the batch culture, a maximal cell density of 18.7±0.1 g DW·L^{-1} in the flask was obtained after 14 days of fermentation. In the fed-batch culture, a maximal cell density of 22.9±1.2 g DW·L^{-1} was reached after 14 days of fermentation. The results indicated that lactose feeding was effective to enhance the cell density.

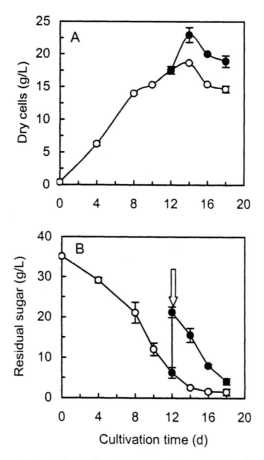

Fig. 3. Time profiles of dry cell weight (A) and residual medium sugar (B) in batch and fed-batch fermentations of G. lucidum in shake flasks. In the figure, the arrow indicates the lactose feeding point. Symbols: batch, open symbol; fed-batch, dark symbol. The error bars in the figure indicate the standard deviations calculated from 2-3 independent samples.

The cell density obtained in this work was much higher than that in previous reports (7,8,11). Lactose addition also led to an increase in β-galactosidase activity of *G. lucidum* cells growing on lactose (data not shown). *G. lucidum* was demonstrated to possess the activity of β-galactosidase, which implies the existence of a lactose permease system for lactose transport into the higher fungus.

Enhanced α-PGM Activity and EPS Production by Lactose Feeding

Kinetics of α-PGM activity in batch and fed-batch fermentations of *G. lucidum* were compared in Fig. 4A. During batch culture, the maximum activity of α-PGM was 1.65 ± 0.06 μmol NADH (mg protein)$^{-1}$ min^{-1} on day 16. Lactose addition caused an immediate increase of α-PGM activity. During fed-batch culture, the maximum α-PGM activity was 1.93 ± 0.06 μmol NADH (mg protein)$^{-1}$ min^{-1} on day 16. The EPS accumulation pattern was shown in Fig. 4B. During batch culture, EPS production titer and productivity was 0.67 ± 0.04 g·L^{-1} (day 16) and 40.6 ± 2.5 mg·(L·d)$^{-1}$, respectively, while they were 1.10 ± 0.02 g·L^{-1} (day 16) and 67.5 ± 1.3 mg·(L·d)$^{-1}$ in the fed-batch cultivation. A remarkable improvement of both the production and productivity of EPS was achieved by lactose feeding. The above results suggest that there may exist correlation between α-PGM activity and EPS biosynthesis by *G. lucidum*, and lactose addition may be useful to regulate the metabolism and physiology of the medicinal fungus.

Responses of PGI Activity and Lactate Accumulation to Lactose Addition

Kinetics of PGI activity and lactate concentration in batch and fed-batch cultures of *G. lucidum* were compared in Figs. 5A and 5B, respectively. During batch fermentation, a maximal PGI activity of 4.97 ± 0.31 μmol NADH (mg protein)$^{-1}$ min^{-1} was obtained on day 16, and a maximal lactate concentration of 73.7 ± 4.5 mg·L^{-1} was obtained after 12 days fermentation. The PGI activity was immediately increased after lactose feed, and a maximal PGI activity of 7.31 ± 0.16 μmol NADH (mg protein)$^{-1}$ min^{-1} was reached on day 16.

Concluding Remarks

A higher cell growth rate and a higher productivity of IPS and GA were reached under higher O_2 supply. The positive response of α-PGM activity and EPS accumulation by *G. lucidum* to sugar feeding was demonstrated. A remarkable improvement of both the production and productivity of EPS was successfully achieved by lactose feeding. The results are considered useful for the regulation and optimization of the mushroom cultures for simultaneous,

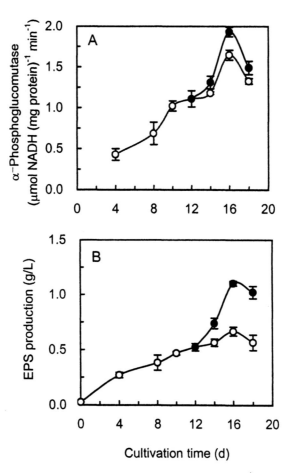

Fig. 4. Time courses of α-phosphoglucomutase (α-PGM) activity (A) and EPS production (B) in batch and fed-batch fermentations of G. lucidum. *The symbols are the same as in Fig. 3.*

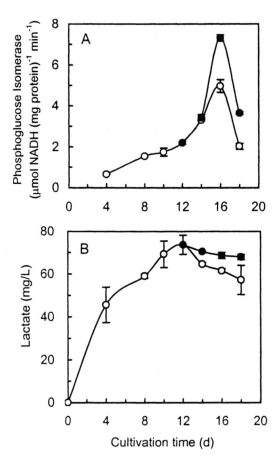

Fig. 5. Kinetics of phosphoglucose isomerase (PGI) activity (A) and lactate concentration in fermentation broth (B) in batch and fed-batch fermentations of G. lucidum. The symbols are the same as in Fig. 3.

highly efficient production of the cell mass, *Ganoderma* polysaccharide, and ganoderic acid on a large scale.

Acknowledgements

Financial support from the National Natural Science Foundation of China (NSFC project no. 20076011), the Science and Technology Commission of Shanghai Municipality (Qimingxing Jihua), and the Key Discipline Development Project of Shanghai Municipality (Shanghai-shi Zhongdian Xueke Jianshe Xiangmu) is acknowledged. The Cheung Kong Scholars Program administered by the Ministry of Education of China and the National Science Fund for Distinguished Young Scholars administered by NSFC are also appreciated.

References

1 Tang YJ, Zhong JJ (2000) Submerged fermentation of higher fungi for production of valuable metabolites (a review). In: Zhong JJ, ed. *Advances in Applied Biotechnology* (in English). Shanghai: East China University of Science and Technology Press, pp. 104-118.

2 Sone Y, Okuda R, Wada N, Kishida E, Misaki A (1985) Structures and antitumor activities of the polysaccharides isolated from fruiting body and the growing fermentation of mycelium of *Ganoderma lucidum*. Agric. Biol. Chem. 49: 2641-2653.

3 Hikino H, Konno C, Mirin Y, Hayashi T (1985) Isolation and hypoglycemic activity of ganoderans A and B, glycans of *Ganoderma lucidum* fruit bodies. Planta Med. 4: 339-340.

4 Shiao MS, Lee KR, Lee JL, Cheng TW (1994) Natural products and biological activities of the Chinese medicinal fungus *Ganoderma lucidum*. In: *ACS Symp Ser 547 Food Phytochemicals for Cancer Prevention II*, Washington DC: American Chemical Society, pp. 342-354.

5 El-Mekkaway S, R. Meselhy M, Nakamura N, Tezuka Y, Hattori M, Kakiuchi N, Shimotohno K, Kawahata T, Otake T (1998) Anti-HIV-1 and anti-HIV-protease substances from *Ganoderma lucidum*. Phytochemistry 49: 1651-1657.

6 Wu TS, Shi LS, Kuo SC (2001) Cytotoxicity of *Ganoderma lucidum* triterpenes. J Nat Prod 64: 1121-1122.

7 Lee KM, Lee SY, Lee HY (1999) Bistage control of pH for improving exopolysaccharide production from mycelia of *Ganoderma lucidum* in an air-lift fermentor. J. Biosci. Bioeng. 88: 646-650.

8 Yang FC, Liau CB (1998) The influence of environmental conditions on polysaccharide formation by *Ganoderma lucidum* in submerged fermentation. Process Biochem. 33: 547-553.

9 Tsujikura Y, Higuchi T, Miyamoto Y, Sato S (1992) Manufacture of ganoderic acid by fermentation of *Ganoderma lucidum* (in Japanese), Jpn Kokai Tokkyo Koho JP 04304890.

10 Fang QH, Zhong JJ (2002a) Two-stage culture process for improved production of ganoderic acid by liquid fermentation of higher fungus *Ganoderma lucidum*. Biotechnol Prog 18: 51-4.

11 Fang QH, Zhong JJ (2002b). Effect of initial pH on production of ganoderic acid and polysaccharide by submerged fermentation of *Ganoderma lucidum*. Process Biochem 37: 769-774.

12 Fang QH, Zhong JJ (2002c). Submerged fermentation of higher fungus *Ganoderma lucidum* for production of valuable bioactive metabolites - ganoderic acid and polysaccharide. Biochem Eng J 10: 61-65.

13 Tang YJ, Zhong JJ (2002) Fed-batch fermentation of *Ganoderma lucidum* for hyperproduction of polysaccharide and ganoderic acid. Enzyme Microb Technol 31: 20-28.

14 Ishmentskii AA, Kondrat'eva TF, Smut'ko AN (1981) Influence of the acidity of the medium, conditions of aeration and temperature on pullulan biosynthesis polyploid strains of *Pullaria (Aureobasidium) pullulans*. Mikrobiologiya 50: 471-475.

15 Hajjaj H, Blane PJ, Groussac E, Goma G, Uribelarrea JL, Loubiere P (1999) Improvement of red pigment/citrinin production ratio as a function of environmental conditions by *Monascus ruber*. Biotechnol Bioeng 64: 497-501.

16 Yoshida T, Shimizu T, Taguchi H, Teramoto S (1965) Studies of submerged culture of basidiomycetes (II) the effect of oxygen on the respiration of Shiitake (*Lentinus edodes*). J Ferment Technol 43: 901-908.

17 Yoshida T, Shimizu T, Taguchi H, Teramoto H (1967) Studies on submerged cultures of basidiomycetes (III): the oxygen transfer within the pellets of *Lentinus edodes*. J Ferment Technol 45: 1119-1129.

18 Ramos A, Boels IC, Willem M. de Vos, Santos H (2001) Relationship between glycolysis and exopolysaccharide biosynthesis in *Lactococcus lactis*. *Appl. Environ. Microbiol.* 67:33-41.

19 Looijesteijn PJ, Boels IC, Kleerebezem M, Hugenholtz J (1999) Regulation of exopolysaccharide production by *Lactococcus lactis* subsp. *cremoris* by the sugar source. *Appl. Environ. Microbiol.* 65:5003-5008.

20 Hickey MW, Hillier AJ, Jago GR (1986) Transport and metabolism of lactose, glucose, and galactose in homofermentative *Lactobacilli*. *Appl. Environ. Microbiol.* 51:825-831.

21 Wang SJ, Zhong JJ (1996) A novel centrifugal impeller bioreactor. II. Oxygen transfer and power consumption. Biotechnol Bioeng 1996;51:520-527.

22 Bradford MM (1976) A rapid and sensitive method for the quantitation of microgram quantities of protein utilizing the principle of protein-dye binding. *Anal. Biochem.* 72:248-254.

23 Gaspar A, Cosson T, Roques C, Thonart P (1997) Study on the production of a xylanolytic complex from *Penicillium canescens* 10-10c. Appl. Biochem. Biotechnol. 67: 45-58.

24 Randers-Eichhorn L, Bartlett RA, Frey DD, Rao G (1996) Noninvasive oxygen measurements and mass transfer considerations in tissue culture flasks. Biotechnol. Bioeng. 51: 466-478.

Chapter 8

Culture Methods for Mass Production of Ruminant Endothelial Cells

José L. Moreira[1], Pedro M. Miranda[1], Isabel Marcelino[1], Paula M. Alves[1], and Manuel J. T. Carrondo[1,2*]

[1]Instituto de Biologia Experimental e Technologica/Instituto de Technologia Quimica e Biológica, Apartado 12, 2781–901 Oeiras, Portugal
[2]Lab. for Engenharia, Bioquímia, Fac, Ciências e Tecnologia, Universidade Nova de Lisboa, 2825 Monte da Caparica, Portugal

Mass production of finite cultures of goat jugular endothelial cells is required for the manufacturing in a cost effective manner, of an inactivated vaccine against Heartwater, a cattle disease caused by the bacteria *Rickettsia Cowdria ruminantium* that is endemic in Sub-Saharan African countries and West Indies. Two different approaches with potential for industrial purpose were tested to produce large amounts of viable caprine jugular endothelial (CJE) cells: static cultures and stirred tanks. Static cultures were easily scaled-up; similar cell growth patterns and final cell concentrations *per* cm^2 were achieved for a 700 fold increase of the surface area. In these culture systems a critical variable was the inoculum concentration, which should be maintained in the range of 1 to 2×10^4 cells/cm^2. CJE growth was also possible in stirred tanks, after optimization of the most relevant variables. The optimal conditions for mass production were 6 g/L of Cytodex 3 using an inoculum of 3.3×10^5 cells/cm^3 in vessels operated at 40 rpm (Reynolds number of approximately 1.5×10^3). The amount of cells produced in a 250 ml stirred tank was similar to that obtained with a 6320 cm^2 Cell Factory^{TM} reactor. The final decision upon the chosen culture method will depend upon the efficiency of the bacterial infection and vaccine production processes that can be derived from these mass cell growing processes.

Endothelial cells constitute the inner monolayer of blood vessels. Due to their proximity to the flowing blood and tissues, they mediate and regulate a large number of important physiological processes that occur on the wall vessel. Among those are: the formation of the preliminary vascular plexus and the generation of blood vessels during the development of embryos *(1,2)*, blood pressure regulation and vasodilatation *(3,4)*, chronic rejections after surgery *(5,6)*, wound healing *(7)*, immune response *(8,9)* and the transport of several macromolecules into the brain *(10)*; moreover, these cells are involved in cancer metastasis *(11)*.

Endothelial cells are also very relevant for veterinary studies because they can be infected by a bacteria of the *Rickettsia* family, *Cowdria ruminantium*, that lives in the vascular system, is taken up by neutrophils and monocytes of wild and domestic ruminants and is the agent of Heartwater (or Cowdriosis), a tick-borne disease that is endemic in Sub-Saharan Africa and in the West Indies *(12)*. This disease is one of the most economically important tick-borne diseases of ruminants; thus, an effective vaccination of domestic animals in a cost effective manner is essential for the limitation of its economic consequences in less developed countries. Presently, a vaccine based on the inactivated bacteria is the best candidate for veterinary use *(13-15)*. In vitro, *Cowdria ruminantium* is a specific parasite of finite cultures of ruminant endothelial cells *(12)*; consequently, a large-scale production of the vaccine requires the mass production of the host cells. This is a particularly difficult issue because these ruminant endothelial cells have a poor ability to proliferate *in vitro* namely due to their limited number of cell duplications (finite cultures), low specific cell growth rates and maximum cell concentrations, associated with an almost uncontrollable phenotypic variability. Currently there is no method described in the literature for the cultivation of significant amounts of ruminant endothelial cells. Moreover, it is well reported that the culture method used for the growth of several types of endothelial cells is critical, since culture conditions influence their behavior, characteristics and activity. Previous reports have shown that other types of endothelial cells grown under shear stress often present significant changes in cell morphology *(16,17)*, microstructure, size *(6,18)* and biochemical activities *(19)*, eventually leading to cell differentiation.

The most straightforward strategy that completely eliminates the negative effects of shear stress is to culture cells in static conditions. Currently several options are available in the market that allow the scale-up and large-scale cultivation of cells using this methodology, such as roller bottles, Cell Factories™ (Nunc) and CellCube™ (Costar). Nevertheless, given the increased homogeneity, aeration and scalability provided by stirred tanks, these may be the best culture system for large-scale propagation of animal cells. Endothelial cells are anchorage-dependent and require a solid matrix to adhere and grow in stirred conditions. The growth of endothelial cells of human and bovine origin in porous and non-porous microcarriers has been previously described *(20-22)*. From these reports it is clear that the origin of the endothelial cells is critical for the optimization of culture conditions (e.g. microcarrier type and concentration, inoculum concentration and agitation rate) and influences cell growth and maximum cell concentration.

The main goal of this work was the optimization of preliminary techniques and the development of new methodologies for mass production of endothelial cells. Static surfaces and stirred tank bioreactors were tested and optimized for the growth of CJE cells. The different culture methods were compared at both technical and economical levels.

Methodologies

Caprine jugular endothelium cells (CJE), isolated from different goats, were isolated by Dr. Dominique Martinez (CIRAD/EMVT, Guadeloupe, France). Cells were propagated in T-flasks (Nunc, Roskilde, Denmark) in a humidified atmosphere of 7% CO_2 in air at 37°C. Unless stated otherwise, Dulbecco's Modified Eagle's Medium (DMEM) supplemented with 2 mM of glutamine, 10 % of fetal bovine serum (FBS), penicillin/streptomycin (100 U/mL) and fungizone (1% v/v) was used as culture medium (all final concentrations and from Invitrogen, Glasgow, UK). Routinely, cells were subcultured with a dilution ratio of 1:3 or 1:5 in similar T-flasks whenever 100% confluence was achieved. The cell detachment from the surface of the T-flask was done by trypsinization using a 0.2% (w/v) Trypsin/EDTA solution (from Invitrogen). These cells do not form stable cell lines and start to differentiate after 25 to 30 passages. Cell differentiation can be detected by morphologic changes such as increase in cell volume, decrease in the cell growth rate and maximum concentration. To circumvent this drawback, cells were discarded after passage number 20.

Culture Medium Studies

Several media were tested to grow CJE cells: DMEM, HAM's F12, DMEM/HAM's F12 (1:1), RPMI 1640, RPMI 1629 and Glasgow Medium (all from Invitrogen), supplemented with 10 and 20% of FBS. In these studies the cells were grown in the conditions referred previously, and then diluted in the new conditions. These experiments were repeated at least 3 times with similar results.

Static Culture Studies

Scale-up studies in static cultures were performed using the procedure described above for cell propagation and a defined inoculum concentration (stated in the text). In those studies, various static culture systems were used: 6 well plates, tissue culture flasks (all standard sizes from Nunc), Cell Factory™ (from Nunc), and also roller bottles (from Corning Costar, Badhoevedor, The Netherlands). For roller bottle cultures, a period at low rotation (12 rph) was used during the initial 24 hours (to promote optimal cell attachment) after which the rate was increased to 24 rph. Despite the fact that roller-bottles are not conventional static cultures, the results obtained with this culture system were included in this section for a more comprehensive interpretation of the results.

Cell number and viability were evaluated using the trypan blue dye exclusion method (0.4% (w/v) in phosphate buffer saline (PBS)) and a hemacytometer (Brand, Wertheim, Main, Germany) after cell detachment by trypsinization as described above in the cell propagation section (several washing steps using PBS were required for cell detachment in the Cell Factory™).

Stirred tank studies

Stirred tank studies were performed in 250 cm^3 spinner flasks from Wheaton Magna-Flex (Techne, New Jersey, USA). The culture medium was DMEM as described above for cell propagation. Several types of microcarriers were used: Cytodex 2, Cytodex 3 and Cytopore 2 from Pharmacia (Upsalla, Sweden), Cultispher G from Percell (Lund, Sweden) and Biosilon from Nunc. The microcarriers were prepared and sterilized as described by the manufacturers. Unless stated otherwise, an inoculum of 10^5 cells/cm^3 was used in all studies. The inoculation procedure is particularly critical for an efficient cell attachment: cells were left in contact with the microcarriers in half of the bioreactor final volume, for 2 hours, without agitation, followed by 24 hours of smooth agitation (30 to 40 rpm). Immediately after that period the culture medium was adjusted to 250 cm^3 and the agitation rate was set up to the final value. Unless stated otherwise, Cytodex 3 was used at 40 rpm.

The bioreactor sampling was done in two steps: in the first step, samples from the supernatant were collected after microcarrier sedimentation. For that purpose the agitation was arrested for a maximum of 3 min. and the number of non-attached cells was evaluated by the trypan blue dye exclusion method. After that, agitation was restarted and a sample containing microcarriers was collected, the microcarriers were allowed to settle and the culture medium was discarded. After two washing steps with PBS, the microcarriers were trypsinized for cell detachment. Cell number was counted and viability was evaluated using the trypan blue dye exclusion method.

Results and Discussion

Medium Culture

The main goal of this work was the development and optimization of culture methods for mass production of finite cultures of caprine endothelial cells (CJE) that, eventually, will be infected with *Cowdria ruminantium* and used for the production of a vaccine. This final product being intended for a veterinary vaccine, the lowest production cost is a critical factor for the success of the process. Several culture media were tested and their performance for endothelial cell growth evaluated. The results obtained for static cultures (final cell concentration 196 hours after inoculation) are shown in Figure 1.

The best culture medium was DMEM, all other alternatives showed a 15-20% lower performance. Similar results were obtained after serial propagation in each culture media tested (at least 5 dilutions), the cell viability always being higher than 95%. Other media (M199, MCDB, Iscove's and Iscove's/HAM's F12 1:1) were also tested, leading to much lower cell yields than those presented in Figure 1 (data not shown). Our results indicate that the optimal culture medium depends on the origin of the cells, since they are compared with results previously reported in the literature for human umbilical cord vein endothelial

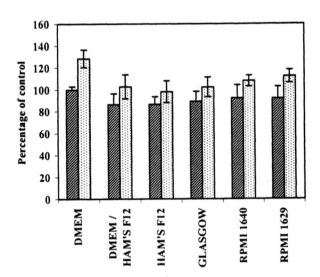

Figure 1. Comparison of different basal media supplemented with 10 (◪) and 20% of FBS (▣) on the growth of CJE cells in static cultures. The results represent the mean of three samples, collected 6 days after inoculation. The control for comparison (100%) is DMEM 10% FBS.

cells (HUVEC's) where a 1:1 mixture of Iscove's MDM/Ham's F12 showed the best growth yields *(23)*.

The serum concentration is a significant parameter to take into account for the production of large amounts of endothelial cells. The use of 20% FBS increased by almost 25% the final cell concentration (Figure 1). However, a final decision will depend upon the cost increase due to the use of 20% of serum: for DMEM and static cultures this increase in the serum concentration leads to a 30% increase in the total final costs over 10% serum. Thus, 20% of serum should not be used unless the yield in the final bacterial infection is very substantially improved. Significant changes on cell morphology and size were observed for the different culture media tested (independent of the serum concentration): cells grown in RPMI 1629 are very thin and long, whereas they look wider in Glasgow and more spread in DMEM. As previously reported by Weiss *et al. (25)*, this could not be correlated with cell metabolic activity or growth patterns, since similar morphologies were observed for DMEM and for serum free medium where no cell growth was detected.

One important observation is that in our experiments the cells grew up to concentrations expected for endothelial cells *(24)* without any addition of expensive medium supplements such as fibroblast or epidermal growth factors, normally used for the culture of most endothelial cells *(16,26)*.

Cell Growth in Static Cultures

In order to define and optimize the best conditions for mass production of CJE cells on static cultures, two of the best out of six batches of cells were chosen after preliminary testing (CJE 102 and CJE 301). A characterization of the cell growth curves was performed using 6 well plates. The effect of serum concentration (10 and 20%) was also evaluated. For this purpose, an inoculum of 0.5×10^4 cells/cm^2 was plated and the cell concentration evaluated along time. The two cell batches show similar growth patterns for both serum concentrations tested (Figure 2A and Figure 2B). Despite the fact that a maximum absolute cell concentration was achieved for CJE 301 cells, approximately 1.2 times higher than for CJE 102 (Figure 2), which can be explained by the smaller size of CJE 301 cells, a compact monolayer was observed in both cases and no significant batch to batch variability was found.

The data presented in Figure 2 is in close agreement with results previously reported for endothelial cells of other origins such as HUVEC's (24). Nevertheless, the specific growth rate (0.012 to 0.017 h^{-1} obtained, leading to duplication times of 40 to 60 hours) and the final cell concentrations (5 to 6 x 10^4 cells/cm^2 with 10% FBS) are lower than those normally achieved for other mammalian cells, e.g. fibroblasts.

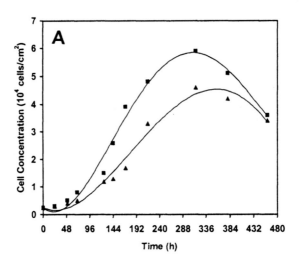

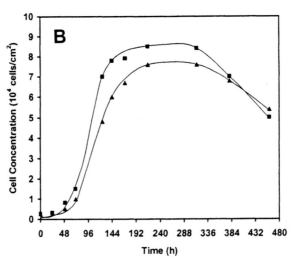

Figure 2. Growth of CJE 102 (▲) and CJE 301 (■) cells in static cultures, with DMEM supplemented with 10% (A) and 20% (B) FBS.

Scale Up of Static Cultures

To evaluate the effect of the available surface area in CJE growth in static cultures, a large range of areas for growth were studied, from 9.6 cm^2 up to 6 320 cm^2 (Cell Factory^{TM} with 10 plates). The most critical variable to be optimized was the inoculum concentration. Low inoculum requires longer lag phases and an increase in the total time of a production cycle, whereas a high inoculum implicates larger amounts of cells to be seeded, thus higher costs and lower cell yields. The inoculum concentration was varied from 0.1 to 3 x 10^4 cells/cm^2 in 6 well plates (9.6 cm^2); afterwards, the time to confluence, corresponding to a compact monolayer with a cell concentration of 5 to 6 x 10^4 cells/cm^2, was monitored. The results are shown in Table I. As expected, when the inoculum concentration increased the time required for cell confluence decreased, varying from 4 to 12 days. The optimal relationship between maximum cell yield and minimum time of a production cycle was 1 to 2 x 10^4 cells/cm^2. Similar results were obtained with studies performed in 500 cm^2 triple T-flasks (data not shown).

Table I. Influence of inoculum concentration on the time to confluence of CJE 102 cells in static cultures

Inoculum concentration (10^4 cells / cm^2)	Time to 100% confluence (days)
0.1	12
0.5	8
1	7
2	5
3	4

The culture of CJE cells in roller bottles was also evaluated using 900 cm^2 surfaces. The roller bottles were operated at a rotation rate of 24 rph using 100 to 150 ml of medium. For this system the optimal inoculum concentration was also 2 x 10^4 cells/cm^2 (confirming the results presented in Table I).

Static cultures with growth areas varying from 9.6 up to 6320 cm^2 were compared with regard to cell growth kinetics and final cell concentration. The results (Figure 3 and Table II) show that cell growth is independent upon scale the final cell concentration at confluence being, on average, 5.4 x 10^{4}cells/cm^2. The cell viability was always higher than 95%. Figure 3 also shows that the cell growth pattern was independent from the size of T-flasks (growth rate of 0.012 h^{-1}). In the larger culture system (Cell Factory^{TM}), 3.6 x 10^8 cells were obtained, which is a significant amount of cells that might justify the possibility of using it as a production system for the Heartwater vaccine.

One of the major achievements of this work was the mass production of finite cultures of endothelial cells; our experiments show that the cell growth pattern is independent of the area available for growth thus, larger static culture systems (as CellCube^{TM}) can be proposed, if economically advantageous. There are several industrial processes based on roller bottles and this system can be regarded as an alternative production process. Nevertheless, roller bottles have potential disadvantages for the production of this vaccine because they require large controlled rooms for cell growth, specific equipment and substantial handling.

Table II. Mass production of CJE cells: scale-up in different types of static cultures

Culture method	Growth area (cm²)	Total final cell number (10⁶ cells)	Final cell number per unit of area (10⁴ cells/cm²)
6 well plate	9.6	0.5	4.7
	25	1.6	6.2
T-flask	75	3.8	5.1
	175	10.0	5.7
Triple T-flask	500	23.5	4.7
Roller bottle	900	52.0	5.8
Cell Factory^{TM}	6320	360.0	5.7

Stirred Tank Cultures Using Microcarriers

Stirred tanks constitute the most convenient culture method for the majority of the biological systems, due to their simplicity, easy control and scale-up, low cost, and well known operation. One of the major drawbacks on the use of stirred tanks for animal cell growth is the shear stress, which is particularly critical when microcarriers are required for cell attachment (as for CJE cells). The most relevant parameters to optimize in these culture systems are related with the microcarriers characteristics (type, concentration and inoculation procedures) and the conditions that lead to an efficient supply of oxygen to the cells under the minimum shear stress (agitation rate and aeration). In this study those variables were evaluated and optimized.

The attachment of the cells to a surface is a critical step for cell growth, thus the type of microcarrier used is very relevant. Several types of supports were tested: non-porous (Cytodex 2, Cytodex 3 and Biosilon) and macroporous microcarriers (Cultispher G and Cytopore 2) even though, for diffusion of bacterial cells for infection this last option does not look favorable. Spinner flasks with a working volume of 250 ml were used; the vessels were agitated at 40 rpm and inoculated with 10^5 cells/cm³. The microcarrier concentration used in each test was 3 g/L for Cytodex 2 and Cytodex 3, 1 g/L for Cultispher G and

Cytopore 2 and 24 g/L for Biosilon (which has a larger particle diameter). These microcarrier concentrations were chosen according to the supplier's recommendations after some adjustments that allowed a direct comparison between the different experiments in terms of available surface *per* cell. The results are shown in Figure 4. The best support for the growth of CJE cells was Cytodex 3 (a non-porous microcarrier), followed by Cultispher G (a porous microcarrier). Despite the fact that Biosilon allowed the cells to grow immediately after inoculation no significant increase of cell number was achieved (max. cell conc. approx. 2×10^5 cells/cm^3) with this microcarrier. All the other supports tested were ineffective for cell proliferation. The two best microcarriers clearly indicate that the immobilization method is not primarily associated with the presence of pores (protection from shear stress) but with the type of surface for cell adhesion.

During cell growth and even before total confluence of the microcarriers, a process of microcarrier aggregation occurred. This phenomenon is well described for anchorage-dependent mammalian cells (27). Moreover, clumps of 10-15 microcarriers containing endothelial cells were previously reported for human umbilical cord vein endothelial cells growing on gelibeads *(24)* and for bovine corneal endothelial cells *(21)*. In this work the clump formation mechanism was similar to that reported in Goetghebeur and Hu *(27)*, especially for the non-porous microcarriers. After initial adhesion, the CJE cells show the expected morphology as they were well spread and elongated on the surface of the support (as observed in static conditions). After 2-3 days the cells start moving to a central point of the microcarrier, grouping themselves in that specific region, forming a multilayer. This process was followed by the clump formation, due to collision of microcarriers and the physical connection between the two carriers on the regions containing cells. In some cases, the release from the supports and the formation of spherical cell aggregates followed this process. After detachment, the cells died, clearly indicating that aggregate formation should be avoided. Nevertheless, by optimization of the relationship between inoculum and support concentrations, the clump formation can be eliminated as will be described below.

The lower performance of Cultispher G, when compared with Cytodex 3, was expected since in this experiment cell growth was not limited by the surface available for cell adhesion. Cultispher is a porous microcarrier thus, it is more difficult for the cells to migrate inside its porous matrix (for an efficient attachment and growth) than to attach immediately after inoculation on the external surface of Cytodex 3. At the agitation rate used in these assays, 40 rpm (corresponding to a Reynolds number of 1.5×10^3 *(28)*), no significant growth limitations due to shear stress on the external surface of Cytodex 3 (approximately 0.027 Nm^{-2} *(28)*) were observed. The maximum cell concentration achieved with the best immobilization method (Cytodex 3) was

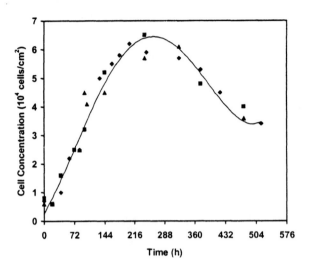

Figure 3. Growth of CJE 102 cells with DMEM supplemented with 10% FBS in static cultures of different sizes: 25 (◆), 75 (■) and 175 (▲) cm².

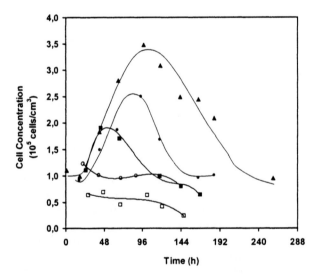

Figure 4. Study of the influence of the support for immobilization of CJE cells in stirred tank batch cultures: Cytodex 3 (▲), Cultispher G (●), Cytodex 2 (O), Citopore 2 (□) and Biossilon (■).

rather low (3.2 x 10^5 cells/cm^3), clearly indicating that this culture system requires further optimization for mass production of endothelial cells.

Effect of Agitation Rate

Another critical parameter for the growth of CJE cells on stirred tanks is the agitation rate. Agitation rates varying from 40 to 100 rpm in 250 ml spinner flasks containing 3 g/L of Cytodex 3 and inoculated with 10^5 cells/cm^3 were tested. The results presented in Figure 5 clearly show that the optimal agitation rate is 40 rpm. At 60 rpm the growth rate was similar to that obtained at 40 rpm (0.020 h^{-1}), but the maximum cell concentration decreased by approximately 30%. This can be due to the faster formation of clusters at 60 rpm due to the increased number of collisions between microcarriers. It is also possible that at 60 rpm the shear stress caused by the agitation rate affected CJE viability: it has been reported before that endothelial cells are very sensitive to shear stress, with immediate changes in cell morphology and growth patterns *(16,17)*.

Agitation rates below 40 rpm can not be used due to the sedimentation of the microcarriers and because they lead to a bad homogeneity in the vessel, particularly for oxygen transfer.

Optimization of Microcarrier and Inoculum Concentrations

As previously reported, in order to use stirred tanks for the mass production of endothelial cells, several problems need to be solved. It is particularly critical to ensure that the surface of the microcarriers is completely confluent and that clump formation is negligible. This can be achieved by the optimization of the support and inoculum concentrations and the relationship between those two parameters. In fact, the requirement of a minimum cell density for a given microcarrier during inoculation could be caused by the need for the right microenvironment on the surface of microcarriers *(29)*.

To assess this issue two groups of experiments were performed using 250 ml spinner flasks agitated at 40 rpm and containing Cytodex 3: in the first group, the inoculum concentration was maintained at 10^5 cells/cm^3 and the microcarrier concentration varied between 0.5 and 5 g/L. In the second group the microcarrier concentration was maintained at 3g/L and the inoculum varied between 0.2 and 5 x 10^5 cells/cm^3. The ranges of both variables were chosen accordingly to the suppliers of the microcarriers and to data previously reported in the literature for other cell types *(30)*. For an easier comparison between the two groups the results shown in Figure 6A and 6B are expressed in terms of expansion factor (ratio between the maximum cell concentration obtained *per* inoculum

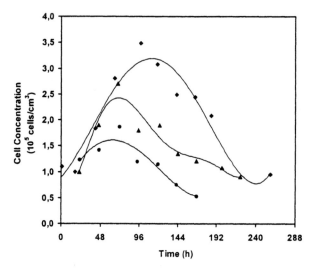

Figure 5. Study of the influence of the agitation rate on CJE cell growth in stirred tank batch cultures on the surface of Cytodex 3: 40 (♦), 60 (▲) and 100 rpm (●).

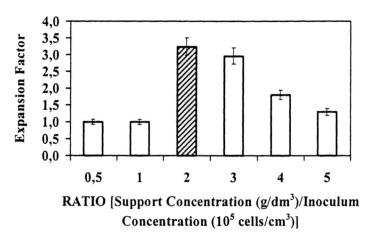

RATIO [Support Concentration (g/dm³)/Inoculum Concentration (10⁵ cells/cm³)]

Figure 6. Optimization of the relationship between inoculum and Cytodex 3 concentrations, by: the variation of the support concentration at 10^5 cells/cm³ of inoculum (A) and the variation of the inoculum concentration at 3 g/L of support (B).

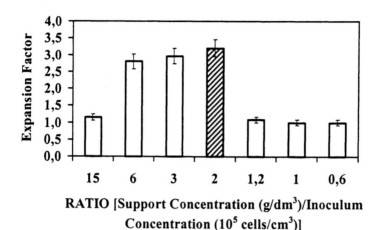

RATIO [Support Concentration (g/dm³)/Inoculum
Concentration (10⁵ cells/cm³)]

Figure 6 Continued

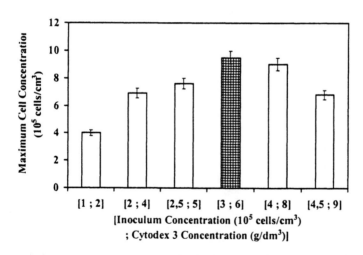

[Inoculum Concentration (10⁵ cells/cm³)
; Cytodex 3 Concentration (g/dm³)]

*Figure 7. Final optimization of the inoculum and Cytodex 3 concentrations for
the growth of CJE cells: variation of the inoculum and support concentrations
maintaining the relationship of 2 grams of Cytodex 3 per 10⁵ cells/cm³ of
inoculum.*

concentration) and ratio of support concentration *per* inoculum concentration. The results demonstrate that the optimal relationship is 2 grams of Cytodex 3 *per* each 10^5 cells/cm^3 of inoculum. From these experiments it can be concluded that the conditions used in the previous studies were sub-optimal, which can justify the low final cell concentrations achieved.

An additional experiment was performed where the ratio 2 grams of Cytodex 3 *per* each 10^5 cells/cm^3 of inoculum was maintained and the Cytodex 3 concentration was varied between 2 and 9 g/L. The results show that the optimal concentrations were: 6 g/L of Cytodex 3 with an inoculum of 3 x 10^5 cells/cm^3 (Figure 7). Approximately 10^6 cells/cm^3 were obtained at the end of the run with the specific growth rate reaching 0.020 h^{-1} and cell viabilities always above 95%. This cell concentration is within the same magnitude of the ones regularly achieved for most mammalian cells grown in stirred tanks operated as batch cultures. As expected, almost all supports were confluent and the clump formation process was significantly reduced. The lower cell yield observed when support concentration is reduced and the ratio of 2 grams of Cytodex 3 *per* each 10^5 cells/cm^3 of inoculum maintained can be due to the effect of the low inoculum size, that requires longer lag phases and/or to an increase in cluster formation, that affects cell viability and growth. The decrease observed at the highest support concentrations tested can be associated with the increase in the number of collisions between supports resulting in increased cell death, already reported for mammalian cells *(30)*. Overall our data shows that the optimal conditions for mass production of endothelial cells in stirred tanks: 6 g/L of Cytodex 3 with an inoculum of 3.3 x 10^5 cells/cm^3 in vessels operated at 40 rpm using DMEM supplemented with 10% FBS being the maximum cell concentration achieved 10^6 cells/cm^3.

Culture Methods Comparison and Cost Estimations

As previously shown (Figure 6 and 7), the stirred tanks operated at the optimal conditions will lead to high cell concentrations. When compared with the static cultures, seven 250 ml stirred tanks could produced 1.75 x 10^9 cells, roughly corresponding to five 6320 cm^2 Cell Factories^{TM} and 25 roller bottles (scaling-up the static culture conditions, this approximately corresponds to the production of one CellCube^{TM} bioreactor). The final method of culture will depend on the efficiency of infection with the bacteria and final amount of vaccine produced, as well as the production cost of the chosen method of culture, which will change substantially when performed in the rich northern or poorer southern hemispheres.

A preliminary cost estimation *per* million of cells for the different culture methods optimized in this work and also for the CellCube^{TM} under the cost constraints prevailing in Europe is presented in Table III. In this estimation the costs of medium and inoculum preparation were also included, since their influence on the final cost will vary with the culture method. The less expensive culture method is the stirred tank reactors, closely followed by roller bottles and by Cell Factories™. Investment costs were not included in this preliminary economic evaluation, which will depend upon the existing facility, structure and equipment for production. Considering that this cost estimations led to relatively close results, the final decision upon the chosen culture method for mass production of CJE cells and Heartwater vaccine will be postponed to include the comparison of the bacterial infection and vaccine production processes and optimization as well as the final location chosen for production.

Table III. Cost estimation of unit of CJE cells produced in different types of cultures.

Culture method	Total cost (Euro)/10^6 cells
Roller bottle	0.35
Cell Factory™	0.41
Stirred tank	0.32
CellCube™	0.52

Conclusions

Our results indicate that the best culture medium for CJE cells is DMEM supplemented with 10% FBS; the use of 20% of FBS leads to an higher cell concentration, but the marginal increase in cell concentration is not compensated by the marginal increase in cost, a critical decision matter in this work geared towards veterinary vaccines. Our data demonstrates that the cell growth pattern is independent upon the animal origin of the CJE cells being the optimal inoculum concentration 1 to 2 x 10^4 cells/cm^2. The cell growth pattern was also independent upon scale-up of support area. This constitutes one of the major findings of this work; large amounts of cells can be produced and used in biological studies or in the production of large amounts of relevant compounds in the absence of shear stress.

We demonstrate for the first time that CJE cell growth is possible in stirred tanks. The optimal conditions for mass production are: 6 g/L of Cytodex with an inoculum of 3.3 x 10^5 cells/cm^3 in vessels operated at 40 rpm, the Reynolds number being approximately 1.5×10^3 and the shear stress 0.027 Nm^{-2}.

Finally we show that stirred tanks, Cell Factories™ and roller bottles can be used for mass production of endothelial cells. A preliminary cost estimation *per* unit of cells identified the stirred tanks as the least expensive culture method, closely followed by roller bottles and Cell Factories™. The final decision upon the chosen culture method will depend upon the efficiency of the bacterial infection, vaccine production processes and location (northern or southern hemisphere).

Acknowledgements

The authors acknowledge and appreciate the financial support received from the European Commission (DGXII/D/3 CTT 634). The authors are also grateful to Dr. Dominique Martinez (CIRAD/EMVT, Guadeloupe) for supplying the cells. Mrs. Cristina Peixoto and Mrs. Maria do Rosário Clemente (IBET) are acknowledged for technical support and Ms. Ana V. Carvalhal for helpful discussions.

References

1. Folkman, J.; Shing, Y. *J. Biol. Chem.* **1992**, *267*, 10931-10934.
2. Rissau, W. *Nature* **1997**, *386*, 671-674.
3. Burnstock, G.; Kennedy, C. *Cir. Res.* **1986**, *58*, 319-330.
4. Snyder, S. H. *Nature* **1995**, *377*, 196-197.
5. Orosz, C. G. *Clin. Transplant.* **1994**, *8*, 299-303.
6. Ott, M. J.; Ballermann, B. J. *Surgery* **1995**, *117*, 334-339.
7. Heimark, R. L.; Twardzik, D. R.; Schwartz, S. M. *Science* **1986**, *233*, 1078-1080.
8. Koch, A. E.; Halloran, M. M.; Haskell, C. J.; Shah, M. R.; Polverini, P. J. *Nature* **1995**, *376*, 517-519.
9. Lobb, R. R.; Chi-Rosso, G.; Leone, D. R.; Rosa, M. D.; Bixler, S.; Newman, B. M.; Luhowskyj, S.; Benjamin, C. D.; Douglas, I. G.; Goelz, S. E.; Hession, C.; Chow, P. *J. Immunol.* **1991**, *147*, 124-129.
10. Rubin, L. L.; Hall, D. E.; Porter, S.; Barbu, K.; Cannon, C.; Horner, H. C.; Janatpour, M.; Liaw, C. W.; Manning, K.; Morales, J.; Tanner, L. I.; Tomaselli, K. J.; Bard, F. *J. Cell Biol.* **1991**, *115*, 1725-1735.
11. Miyake, K.; Yamamoto, S.; Iijima, S. *Cytotechnology* **1996**, *22*, 205-209.
12. Bezuidenhout, J. D. *Onderstepoort J vet Res* **1987**, *54*, 205-210.
13. Martinez, D.; Maillard, J. C.; Coisne, S.; Sheikboudou, C.; Bensaid, A. *Vet. Immunol. Immunop.* **1994**, *41*, 153-163.
14. Totté, P.; McKeever, D.; Martinez, D.; Bensaid, A. *Infect. Immun.* **1997**, *65*, 236-241.

15. Mahan, S. M.; Kumbula, D.; Burridge, M. J.; Barbet, A. F. *Vaccine* **1998**, *16*, 1203-1211.
16. Ballermann, B. J.; Ott, M. J. *Blood Purificat.* **1995**, *13*, 125-134.
17. Barbee, K. A. *Biochem. Cell Biol.* **1995**, *73*, 501-505.
18. Ott, M. J.; Olson, J. L.; Ballermann, B. J. Chronic *In vitro* flow promotes ultrastructural differentiation of endothelial cells. *Endothelium* **1995**, *3*, 21-30.
19. Levesque, M. J.; Sprague, E. A.; Schwartz, C. J.; Nerem, R. M. *Biotech. Progr.* **1989**, *5*, 1-8.
20. Schrimpf, G.; Friedl, P. *Cytotechnology* **1993**, *13*, 203-211.
21. Francis, K. M.; O'Connor, K. C.; Blake, D. A.; Caldwell, D. R.; Spaulding, G. F. In *Animal Cell Technology Developments Towards the 21st century;* Beuvery, E. C., Griffiths, J. B., Zeijlemaker, W. P., Eds.; Kluwer Academic Publishers: Dordrecht, 1995; pp 959-963.
22. Müthing, J.; Duvar, S.; Nerger, S.; Büntemeyer, H.; Lehmann, J. *Cytotechnology* **1996**, *18*, 193-206.
23. Maciag, T.; Hoover, G. A.; Stemerman, M. B.; Weinstein, R. *J. Cell Biol.* **1981**, *91*, 420-426.
24. Friedl, P.; Tatje, D.; Czpla, R. *Cytotechnology* **1989**, *2*, 171-179.
25. Weiss, T. L.; Selleck, S. E.; Reusch, M.; Wintroub, B. U. *In Vitro Cell Dev Biol* **1990**, *26*, 759--768.
26. Gospodarowicz, D.; Cheng, J.; Lirette, M. *J. Cell Biol.* **1983**, *97*, 1677-1685.
27. Goetghebeur, S.; Hu, W. S. *Appl. Microbiol. Biotechnol.* **1991**, *34*, 735-741.
28. Aunins, J. G.; Woodson, B. A.; Hale, T. K.; Wang, D. I. C. *Biotechnol. Bioeng.* **1989**, *34*, 1127-1132.
29. Hu, W. S.; Meier, J.; Wang, D. I. C. *Biotechnol. Bioeng.* **1985**, *27*, 585-595.
30. Cherry, R. S.; Papoutsakis, E. T. *Biotechnol. Bioeng.* **1988**, *32*, 1001-1014.

Chapter 9

High-Density, Defined Media Culture for the Production of *Escherichia coli* Cell Extracts

J. Zawada, B. Richter, E. Huang, E. Lodes, A. Shah,
and J. R. Swartz*

Department of Chemical Engineering, Stanford University,
Stanford, CA 94305–5025

Cell-free protein synthesis systems use a crude cell extract to produce proteins. While these systems have many advantages over *in vivo* expression systems, the current procedures for preparation of the cell extract are very labor-intensive. In addition, since they start with low cell density cultures, the methods yield small amounts of final extract. Fed-batch fermentations were performed with a defined medium and a glucose feeding strategy designed to meet metabolic demand but avoid acetate production. We have investigated cultures with cell densities from 3 to 50 OD_{595} for extract preparation. Furthermore, the effect of growth rate (0.2 to 1.1 /hr) on extract performance was studied in glucose-limited cultures. The glucose feeding strategy was able to control acetate production in some fermentations but not in others. The feeding of amino acids may have caused the variability. Neither cell density nor growth rate had a significant impact on protein synthesis by the cell-free extract. Finally, a *relA* and *spoT* null mutant grew more slowly, but still produced extract with the same activity as the parent *relA1 spoT1* strain.

Introduction

Many limitations of *in vivo* protein expression can be overcome with cell-free synthesis systems including product toxicity, protein folding, and product purification (see reference 1 for a review). However, producing the cell extract used in these systems is an expensive and laborious process. Traditionally, cell extracts are made from batch cultures grown in complex medium and harvested at 3-4 OD_{595}. To improve the yield of extract, we investigated the use of high cell density cultures ($\sim$50 OD_{595}).

One of the main limitations of high cell density *E. coli* cultures is the accumulation of acetate in the medium. When grown on excess glucose, *E. coli* takes up the sugar and processes it through glycolysis faster than it can be handled by the downstream metabolism. The excess pyruvate is converted to acetate and excreted. Eventually, the acetate builds up in the medium and inhibits growth. See Lee's review (*2*) for a discussion of the phenomenon and references to related work. A secondary goal of this work was to develop a glucose feeding strategy that would supply just enough glucose to meet the demand of the culture. Such a method would maximize the growth rate but avoid acetate accumulation.

Several fed-batch fermentation methods have been developed with this goal in mind (see reference 2 for a review); however, they all target a specific growth rate around 0.3 /hr or lower. We wished to grow at a faster rate, and so we judged that careful control of glucose supply with feedback from the culture would be required to sustain growth but avoid acetate production.

Åkesson and co-workers (*3*) used a glucose feeding strategy employing periodic pulses of glucose. The dissolved oxygen response to the pulses was analyzed to determine if glucose was limiting or in excess. The feed rate was then adjusted accordingly. Lin *et al.* (*4*) took the method one step farther. By timing the drop and subsequent rise in dissolved oxygen concentration, they were able to calculate a glucose uptake rate (GUR). As mentioned in the discussion section of their paper, this information could be used to control glucose feeding. This is essentially the approach we used (see Methods section for the details). A similar strategy was used by Takagi *et al.* (*5*).

One additional parameter we wished to examine was growth rate. According to reports in the literature (*6, 7*), ribosome concentration varies with the growth rate. Since ribosomes are obviously critical to any protein synthesis system, we tested the effect of growth rate on extract performance by varying the glucose feed rate.

Materials and Methods

Bacteria Strains

A derivative of *Escherichia coli* A19 was used in all the fermentations. The strain used had the methionine auxotrophy of A19 reverted to prototrophy in addition to the following deletions: Δ*tnaA* Δ*speA* Δ*tonA* Δ*endA*. The strain was constructed in our laboratory. Where noted, a derivative of this strain was used with the Δ*relA251::kan* and Δ*spoT207::cat* alleles (*8*). The alleles were transferred by P1 transduction from the CF1693 strain which was kindly donated by Dr. Michael Cashel, National Institute of Child Health & Human Development.

Defined Medium

A rich defined media was formulated to ensure greater consistency between cultures than is possible with the standard complex media used to grow cells for extract preparation. The media was enriched with many amino acids and vitamins to support rapid growth. The inorganic salt and mineral composition was based on a recipe used industrially for high cell density cultures (*9*). The vitamins were chosen based on chemical analysis sheets obtained from Difco, Inc. which indicated the typical composition of tryptone and yeast extract. All amino acids were supplied except alanine, serine, arginine, cysteine, aspartate, glutamate, and glutamine. More details on the media design will be published elsewhere.

The defined medium used consisted of the following (in g/L): $(NH_4)_2SO_4$ 1.5, $NaH_2PO_4 \cdot H_2O$ 3.5, KCl 2, trisodium citrate dihydrate 0.5. After sterilization, the following vitamins were added (in mg/L): choline chloride 28.6, niacin 25.1, p-aminobenzoic acid (potassium salt) 25.6, pantothenic acid (hemi-calcium salt) 9.4, pyridoxine hydrochloride 1.5, riboflavin 3.9, thiamine hydrochloride 17.7, biotin 0.1, cyanocobalamin 0.01, folic acid dihydrate 0.07. The following minerals were also added after sterilization (in mg/L): $FeCl_3 \cdot 6H_2O$ 20, $Na_2MoO_4 \cdot 2H_2O$ 3.5, boric acid 1.2, $CoCl_2 \cdot 6H_2O$ 3.4, $CuSO_4 \cdot 5H_2O$ 3.4, $MnSO_4 \cdot H_2O$ 1.9, $ZnSO_4 \cdot 7H_2O$ 3.4. The iron(III) chloride and sodium molybdate were prepared in separate stocks with citrate to prevent precipitation. Glucose, magnesium, and amino acids were also added after sterilization (see below).

The glucose feed solution contained 250 g/L glucose and 6.33 g/L $MgSO_4 \cdot 7H_2O$ and was autoclaved. The amino acid feed solution contained the following (in g/L): Asp·H$_2$O 48.9, Gly 70.6, His HCl·H$_2$O 13.5, Ile 35.1, Leu 36.3, Lys·HCl·H$_2$O 30.8, Met 14.1, Phe 14.1, Pro 46.9, Thr 37.1, Trp 14.3, Tyr 17.3, Val 22.8, betaine 33.1. These concentrations are based on observed amino acid consumption rates. Approximately 180 g/L KOH was necessary to dissolve the amino acids and adjust the pH to 11-11.5. The amino acid solution was filter sterilized prior to use. Fresh glucose/magnesium sulfate and amino acid solutions were prepared for each fermentation.

Inoculum Preparation

Several stock vials of each strain were prepared from an overnight culture in LB media. Stocks with 20% (v/v) glycerol were frozen in liquid nitrogen and stored at -80°C until use. For the inoculum, a stock vial was thawed at room temperature and 1 mL was transferred to 5 mL of LB media in a culture tube. The remainder of the stock vial was discarded. The culture was incubated at 37°C with shaking for 4 hours. Then, 1 mL from the tube was used to inoculate 100 mL of defined media with 4 mL of the glucose feed (10 g/L glucose final concentration) and 1.86 mL of the amino acid solution in a 1 L baffled shake flask. The flask was shaken at 37°C overnight for 12-16 hours. Next, 20 mL from the overnight culture was transferred to 400 mL of fresh media with half the amount of glucose and amino acids as the overnight media. Once the culture was in exponential growth and between 2 and 3 OD$_{595}$ (about 3 hours), the entire culture was used to inoculate the fermentor.

Fermentation Conditions

A B. Braun Biostat C fermentor with 10 L working volume was used. The standard baffles were replaced by custom machined 3 cm wide baffles to improve oxygen transfer, especially at high agitation rates. A side feed port for the glucose feed was manufacturing by drilling a hole through a blind stopper and welding a stainless steel tube (1/8 in. ID) into the hole. The outside end of the tubing was connected to a steam supply so the tube could be sterilized by steam injection during the fermentor sterilization cycle.

After sterilization, the dissolved oxygen (DO2) probe was calibrated with argon and air. The MFCS/Win software supplied by B. Braun logged data and controlled the operating conditions as follows: aeration 10 L/min, temperature 37°C, pressure 1.5 bar (gauge), pH 7.2. Ammonium hydroxide (~2 M) and 5 N sulfuric acid were used to control pH. Agitation was started at 400 RPM and increased manually to maintain the DO2 above 30% air saturation at atmospheric pressure.

Initially, the fermentor contained 10 L of the defined media with 93 mL of the amino acid solution added. The glucose/magnesium sulfate solution was added to a final glucose concentration of 5 g/L (200 mL). BASF Mazu® DF-204 antifoam was added (1-2 mL) to control foaming.

Glucose Feeding Strategy

A laptop computer (Toshiba Portégé 7200CTe) with a National Instruments DAQCard-6024E' and MATLAB 6.1 software with the Data Acquisition Toolbox (The Mathworks, Inc. Natick, MA) was used to control glucose and amino acid feeding. The analog recorder output of the Biostat C was connected to the input of the data acquisition card and the card's analog outputs were used to control two MasterFlex 77521-40 peristaltic pumps with EasyLoad II pump heads (Cole-Parmer, Vernon Hills, IL). The DO2 signal from the Biostat C (1 Hz sampling rate) was analyzed by a MATLAB program to detect sharp increases or drops. The point-to-point differences in the most recent 5 samples were averaged and compared to the averaged slope for the previous 85 samples (5-89 seconds ago). If the difference in the two slopes was greater than a threshold value, a sudden rise or drop in DO2 was signaled.

Glucose feeding was started 3.5 hrs after inoculation at a constant 0.6 mL/min to avoid a period of complete glucose exhaustion at the transition from batch to feeding mode. Glucose limitation was detected by a sharp spike in the DO2 which usually occurred between 4 and 4.5 hours of fermentation time. Amino acid feeding was started at 4.25 hrs or when glucose became limiting, whichever came first. Amino acids were fed at 27% of the glucose feed rate, except during glucose pulses when the amino acids feed was maintained on the previous smooth exponential feed profile.

Once glucose limitation was detected, glucose pulse testing started. The glucose feed rate was increased temporarily to supply excess glucose and then reduced until a rise in DO2 was detected. The glucose feed changes were scaled to the current feed rate as follows: feed below 2.5 mL/min, up 75%, down 50%; feed between 2.5 and 5 mL/min, up 50%, down 25%; feed above 5 mL/min, up 30%, down 25%. If glucose was limiting, the pulse of glucose would cause the DO2 to drop as metabolism sped up. Once the excess glucose was depleted, metabolism would slow and DO2 would rise. If a pulse failed to drop the DO2, glucose was assumed to be in excess. The reduced glucose feed rate was maintained until a rise in DO2 indicated glucose limitation. The pulse test was then repeated. The total glucose added during the pulse test was divided by the time from pulse start to DO2 rise to determine the glucose utilization rate. Glucose feeding was then resumed at a feed rate set to 90-100% of the calculated uptake rate. The feed rate was continuously increased exponentially between pulse tests with a rate constant between 0.3 and 1.0 /hr depending on the fermentation (see Results for details). Pulse tests were initiated every 15-20 minutes to adjust the feed rate as necessary.

Sampling

Samples were withdrawn from the side sampling port on the fermentor and passed through a cooling coil in an ice-water bath before being collected in ice cold 15 mL disposable tubes. Approximately 40 mL of culture was flushed through the sampling system before a sample of about 10 mL was collected. A 1 mL portion of the sample was immediately centrifuged at 20,000 g, 4°C for 5 min. Two 0.4 mL portions of the supernatant were frozen in liquid nitrogen and stored at -80°C until later analysis. The cell density was measured at 595 nm by diluting the sample as necessary to get a reading between 0.1 and 0.25 AU.

Glucose and Acetate Chromatography

Glucose and acetate in the sample supernatants were quantitated using an Aminex HPX-87H HPLC column (Bio-Rad, Hercules, CA) on an Agilent 1100 HPLC with a refractive index detector. An isocratic elution was performed with 5 mM H_2SO_4 at 0.4 mL/min and 35°C for 40 min. Injections were 20 μL of undiluted sample supernatant. A calibration curve was prepared with standards from 0.1 to 100 mM. The minimum detectable concentrations were about 0.1 mM for acetate and 0.05 mM for glucose.

Inorganic Phosphate Assay

Inorganic phosphate in the sample supernatants was assayed using the procedure of Saheki et al. (10). Samples were diluted to be in the range of standards (0.05-2 mM) and assayed in triplicate.

Extract Preparation and Protein Synthesis

Extract preparation and protein synthesis of chloramphenicol acetyl transferase (CAT) were done as previously described (11). The extracts designated "2YTPG" in Figure 9 were prepared from cells grown in the traditional manner (9) except that 18 g/L glucose, 3.0 g/L KH_2PO_4, and 7.0 g/L K_2HPO_4 were added to the growth media as described in (12). We have seen no effect of the glucose and phosphate addition on CAT production in this system (unpublished data).

Results

Growth and Acetate Production

The glucose pulse testing strategy described above was first tested with the *E. coli* A19 met[+] Δ*tnaA* Δ*speA* Δ*tonA* Δ*endA* organism with the exponential feed rate constant set at 1.0 /hr.   Figure 1 shows that the growth rate was approximately 1.1 /hr during the batch phase. However, it dropped to about 0.8 /hr during the feeding phase. This is significantly lower than the setpoint of 1.0 /hr used for the glucose feeding. Figure 2 shows that acetate accumulated during the batch phase, as expected. However, once glucose became limiting (about 4.5 hrs), acetate was consumed. Even though the concentration of acetate never exceeded 30 mM, it may have been high enough to inhibit growth and result in the lower than expected growth rate.

Another possible cause of the slow growth is a mild stringent response. The stringent response is a general stress response that can be triggered by several mechanisms including a deficiency in charged tRNAs (mediated by RelA). Although the strain used in the fermentation has mutations in both *relA* and *spoT* (the primary genes involved in producing the stringent response), the *relA1* allele that it carries has been shown to have residual activity (*13*). This activity is significant when combined with a mutant *spoT* allele (*8*) which is the case in our strain. The periodic fluctuations in glucose feeding during the pulse tests may cause temporary starvation for charged tRNAs by restricting energy supply or exhausting metabolic intermediates required for the biosynthesis of amino acids which are not supplied in the medium. To test this hypothesis, alleles which contain large deletions and have no residual activity, Δ*relA251::kan* and Δ*spoT207::cat* (*8*), were transferred from the CF1693 strain into our strain by P1 transduction.

Figure 3 shows that the Δ*relA* Δ*spoT* strain grows more slowly during the batch phase than the parent strain (0.85 /hr versus 1.1 /hr growth rate).   In addition, there still is a drop in growth rate during the feeding phase. Even more troubling is the constant acetate accumulation shown in Figure 4 even though glucose is maintained at low levels during the feeding phase.

Because of the slow growth and acetate accumulation, the relaxed strain was abandoned. The original strain was tested with a reduced feeding rate in an attempt to compensate for the lower growth rate observed during glucose

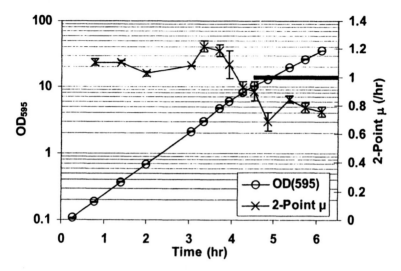

Figure 1. Cell-density (o) and growth rate (x) with 1.0 /hr glucose feed rate constant. The bold horizontal line indicates the period and rate constant of glucose feeding.

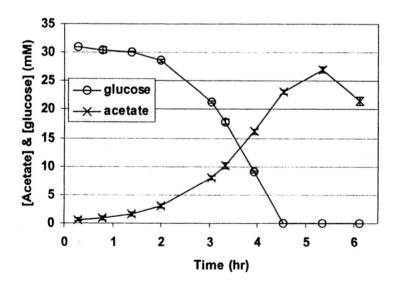

Figure 2. Glucose(o) and acetate (x) levels for the fermentation in Figure 1.

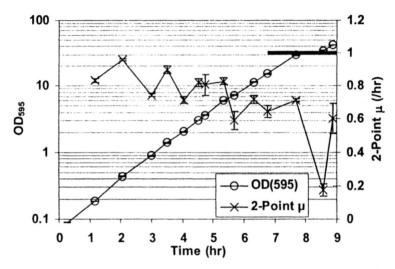

Figure 3. Cell-density (o) and growth rate (x) with 1.0 /hr glucose feed rate constant and the ΔrelA251::kan ΔspoT207::cat mutant. The bold horizontal line indicates the period and rate constant of glucose feeding.

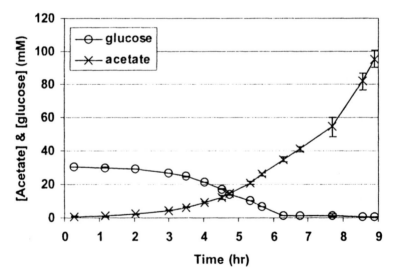

Figure 4. Glucose(o) and acetate (x) levels for the fermentation in Figure 3.

feeding. As shown in Figure 5, the growth rate started at the typical 1.1 /hr during the batch phase, but did not drop to the 0.7 /hr rate of feeding until 1.5 hours after glucose limitation began. Figure 6 shows that glucose was undetectable (<50 μM) from 4.5 hrs on even though the growth rate does not drop to the controlled value until 6 hrs. The acetate concentration also increases throughout the fermentation.

A further reduction in the glucose feed rate produced mixed results. Figure 7 shows that glucose feeding at a rate constant of 0.3 /hr did slow the growth relative to that in the batch phase. But, the growth rate was still significantly higher than the target of 0.3 /hr. Figure 8 shows that glucose was maintained at undetectable levels during the feeding phase until 8 hrs. Also, acetate was not produced during the glucose controlled portion until the very end when glucose rose to detectable levels (0.3-0.4 mM). The increase in glucose concentration is not explained, but it apparently caused the acetate accumulation.

Cell-Free Protein Synthesis

Cell-free extracts were prepared from several samples taken at different points in several fermentations. Chloramphenicol acetyl transferase (CAT) synthesis results are presented in Figure 9. It is clear that neither growth rate nor cell density at the time of cell harvest has a significant impact on CAT production. Furthermore, the extract from the $\Delta relA$ $\Delta spoT$ strain shows the same protein synthetic activity as the other extracts.

Discussion

The defined medium we developed was able to support rapid growth up to 50 OD_{595} and produce extracts with similar performance to that of traditionally prepared extracts. Simply going to a higher cell density will provide a 15-fold improvement in the yield of extract from a fermentation. Preliminary tests indicate that phosphate is limiting the maximum cell density if acetate does not accumulate to high levels. For the fermentation presented in Figures 7 and 8, the final sample contained only 52 ± 1 μM phosphate. Therefore, it should be possible to go to even higher cell densities if the nutrient, specifically phosphate, content of the medium is increased. To avoid precipitation, it may be best to inject more mineral salts at various times or add them to the glucose feed. Eventually, however, the growth rate will need to be constrained to avoid exceeding the available oxygen supply.

Many strategies for glucose feeding in *E. coli* fed-batch fermentations have been reported in the literature (see 2 for a review). While these strategies can

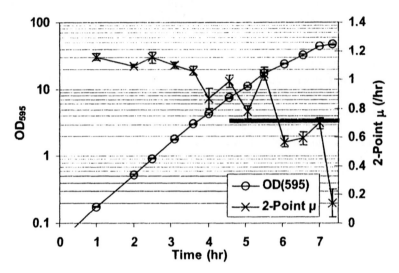

Figure 5. Cell-density (o) and growth rate (x) with 0.7 /hr glucose feed rate constant. The bold horizontal line indicates the period and rate constant of glucose feeding.

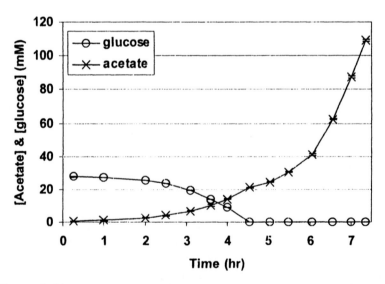

Figure 6. Glucose(o) and acetate (x) levels for the fermentation in Figure 5.

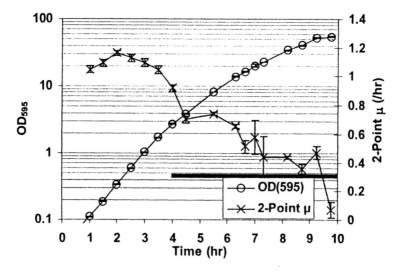

Figure 7. Cell-density (o) and growth rate (x) with 0.3 /hr glucose feed rate constant. The bold horizontal line indicates the period and rate constant of glucose feeding.

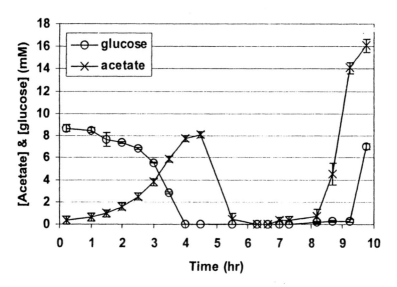

Figure 8. Glucose(o) and acetate (x) levels for the fermentation in Figure 7.

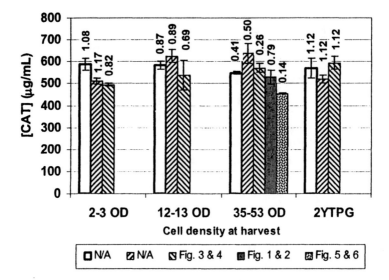

Figure 9. CAT synthesis by extracts from various fermentations. The numbers above the columns indicate the growth rate (in /hr) at the time the cells were collected. The legend indicates which figures contain growth and metabolite data from the fermentation (if applicable). Columns designated "2YTPG" are extracts prepared in the traditional manner (see Methods).

prevent acetate production, they all produce relatively slow specific growth rates (about 0.3 /hr or lower). Our glucose feeding algorithm is designed to support faster growth. Unfortunately, it had mixed results. In some cases glucose and acetate levels were properly controlled. However, the continued acetate accumulation in cases where the glucose concentration was essentially zero indicates there is a problem with the system. Possibly, the glucose levels were actually higher than measured if some glucose was consumed by the cells in the process of sampling and centrifugation. Alternatively, the feeding of amino acids may have caused some acetate production. Han *et al.* (*14*) have shown that glycine (one of the amino acids we feed) can be converted to acetate and excreted. Another possibility in that the pulse tests were not done frequently enough to adjust to the declining glucose uptake capacity that has been reported in some *E. coli* fermentations (*4, 15*).

In the future, the method will be tested without amino acids in the medium or feed. Originally, amino acids were included in the medium to support the highest growth rate. However, from the data presented here, growth rate does not appear to be an important parameter in determining extract productivity. Omitting the amino acids should allow for better growth rate control by glucose feeding in addition to potentially eliminating acetate production during glucose control.

The lack of an impact of growth rate on extract performance is curious. There is clear evidence in the literature (*6, 7*) that ribosome content increases with faster growth rates. Apparently some other component is limiting protein production in our cell-free system and ribosomes are present in saturating quantities. Alternatively, the extract preparation procedure may reduce all extracts to an equivalent activity regardless of the original state of the cells used. Yet another possibility is that in our fermentations the cells have not adjusted to the new, lower growth rates by the time they are harvested.

Acknowledgements

The CF1693 strain containing the $\Delta relA251::kan$ and $\Delta spoT207::cat$ alleles was kindly donated by Dr. Michael Cashel of the National Institute of Child Health & Human Development division of the National Institutes of Health. The assistance of David Liu in extract preparation and testing is greatly appreciated.

References

1. Jewett, M.; Voloshin, A.; J. Swartz; In *Gene Cloning and Expression Technologies;* Weiner, M. P.; Lu, Q., Eds.; Eaton Publishing: Westborough, MA, *in press*.

2. Lee, S. Y.; *Trends Biotechnol.* **1996,** *14,* 98-105.
3. Åkesson, M.; *et al. Biotechnol. Bioeng.* **2001,** *73,* 223-230.
4. Lin, H. Y.; *et al. Biotechnol. Bioeng.* **2001,** *73,* 347-357.
5. Takagi, M.; *et al. Biotechnol. Bioeng.* **1996,** *52,* 653-660.
6. Bremer, H.; Dennis, P. P. In Escherichia coli *and* Salmonella: *cellular and molecular biology,* 2nd ed.; Neidhardt, F. C., Ed. ASM Press: Washington, D.C., 1996; Vol. 2, pp 1553-69.
7. Yun, H. S.; *et al. Biotechnol. Bioeng.* **1996,** *52,* 615-624.
8. Xiao, H.; *et al. J. Biol. Chem.* **1991,** *266,* 5980-90.
9. Swartz, J. R. U.S. Patent 5,633,165, 1997.
10. Saheki, S.; *et al. Anal. Biochem.* **1985,** *148,* 277-281.
11. Kim, D. M.; Swartz, J. R. *Biotechnol. Bioeng.* **2001,** *74,* 309-316.
12. Kim, R. G.; Choi, C. Y. *J. Biotechnol.* **2000,** *84,* 27-32.
13. Metzger, S.; *et al. J. Biol. Chem.* **1989,** *264,* 21146-52.
14. Han, L.; *et al. J. Biotechnol.* **2002,** *92,* 237-249.
15. Konstantinov, K. B.; *et al. J. Ferment. Bioeng.* **1990,** *70,* 253-260.

Environmental Remediation

Chapter 10

Bioprocess Development for Mercury Detoxification and Azo-Dye Decolorization

Jo-Shu Chang

Department of Chemical Engineering, National Cheng Kung University, Tainan, Taiwan

Biological treatment of environmental pollutants is often very complex and requires intensive control of operation parameters. Fermentation of microorganisms able to degrade or destroy pollutants is a key technology to ensure appropriate cell concentration and populations are maintained in the treatment system. Optimal operation strategies, especially policy for substrates feeding and oxygen supply, need to be developed to enhance the rates of biodegradation or biotransformation for efficient cleanup of the target pollutants. In this chapter, two unique types of biological treatment processes, namely, mercury detoxification and azo-dye decolorization, are introduced as the example to show how fermentation technology and bioreactor approaches are used to develop bioprocesses for environmental applications. Detoxification of mercury is conducted aerobically using wild-type and recombinant strains as the biocatalyst. The performance and stability of fed-batch and chemostat bioreactors for mercury detoxification are greatly correlated to the dosage of the toxic mercury substrate. In contrast, azo dye is less toxic but decolorization of azo dyes requires aerobic/anaerobic sequential environments. Thus, bioreactor design for bacterial decolorization focused on strategies for the supply of oxygen and retention of high cell concentration within the reactors to achieve optimal combinantion of aerobic cell growth and anaerobic decolorization. The effectiveness and feasibility of various bioreactor configurations and strategies are assessed.

Introduction

Environmental Biotechnology has become a promising means to clean up pollutants from the contaminated environments in a more natural and cost-effective way. In biological treatment processes, microorganisms capable of degrading or transforming organic or inorganic pollutants into simpler or less toxic forms play a major role.   An efficient biological treatment involves effective growth of biodegraders and control of the bioprocess for an efficient biodegradation or biotransformation of the pollutants. Therefore, fermentation technology is obviously a crucial tool to enable an efficient biological waste treatment process. However, it is a great challenge for fermentation technology when it is applied in waste treatment, since the substrates (pollutants) are more complex and often toxic to the microorganisms as compared with those used in food or pharmaceutical industry.   In addition, the performance of waste degradation is usually very sensitive to the fermentation conditions (pH, aeration, temperature, food to microorganism (F/M) ratio, nutrients, etc.) and may require aerobic/anaerobic sequential environments (*1*) for mineralization of some complex pollutants. For instance, biodegradation of chlorinated aromatic compounds requires anaerobic dechlorination step followed by aerobic ring cleavage of the chlorine-free aromatic compounds and subsequent degradation of the resulting fatty acids via β-oxidation steps (*2*). The other typical example is biological nitrogen removal that converts ammonia to nitrates via an aerobic nitrification step and the nitrates are reduced anaerobically to N_2 with a denitrification step (*3*).   As a result, an appropriate bioprocess design and process control seems to be the key to a successful biological treatment system. This chapter provides examples that demonstrate strategies to develop bioprocesses for mercury detoxification and for azo-dye decolorization. The former is an aerobic process that deals with a toxic substrate. The latter involves an aerobic/anaerobic sequential environment, in which cell growth and degradation of substrate have distinct oxygen requirement.

Bioprocess Development for Microbial Mercury Detoxification

Mercury and mercurial compounds have been used in a variety of industries, causing severe mercury pollution in aquatic systems and soils (*4-6*). Conventional mercury-removal methods mainly via physical and chemical routes to either trap and collect mercury from contaminated sites or to remove mercury by chemical precipitation (*5,7*). The drawback of physical methods has been the requirement for additional treatments, while chemical methods often leave hazardous by-products or residual sludges. Thus, biological methods allowing more natural, efficient and economical cleanup of mercury waste have emerged.

Mercury-resistant microorganisms can resist mercury due to their ability to volatilize soluble forms of mercury from the environment via a sequence of enzymatic reactions, which are recognized as mercury detoxification (*8-11*).

The genetic basis of mercury resistance is encoded in *mer* operons located on either plasmids or transposable elements (12-14). Organomercurial lyase (*mer*B gene product) catalyzes cleavage of C-Hg bonds in organomercurial compounds to release mercuric ions, which are enzymatically reduced to less toxic and more volatile metallic mercury (Hg°) by *mer*A-product mercuric reductase (*12,14-17*). Cysteine-rich transport proteins originated from *mer*P and *mer*T genes locate on the periplasmic space and inner membrane, respectively. The MerP and MerT proteins are responsible for the specific delivery of ambient mercuric ions toward mercuric reductase in the cytoplasm for reduction of Hg^{2+} to volatile Hg° (*16,18*). The constitutive *mer* operon is induced by the subtoxic level of mercuric ions (*13,19*). The mechanism of mercury detoxification is illustrated in Fig. 1.

In contrast to intensive studies and understanding of the molecular genetics of *mer* operon and the biochemistry of mercury detoxification, (*12,14-17*), much less effort has been devoted to develop microbial mercury detoxification processes (*20-22*). Only recently, Saouter et al. (*23*) reported their preliminary investigation of using Hg^{2+}-reducing strains to decontaminate a polluted freshwater pond in Tennessee. Other authors (*24,25*) demonstrated their studies on volatilization of mercury using resting- or immobilized-cell systems.

Recent attempts (*20,26,27*) in our group investigated the kinetic of mercury detoxification by a wild-type mercury-resistant strain (*P. aeruginosa* PU21) and a recombinant *E. coli* strain (*E. coli* PWS1). The two mercury-resistant strains were used to develop batch, fed-batch, and continuous bioreactors to demonstrate operation strategies for efficient and practical detoxification of mercury in wastewater. A mercury vapor recovery device was also designed to recover metallic mercury (Hg°), the end product of mercury detoxification (*26*).

Mercury detoxification occurred when the wild-type or the recombinant strain was cultivated in mercury-containing media. Serial adaptation may be needed to prepare mercury-hyperresistant strain of *P. aeruginosa* PU21 (*26*), while the mercury-reducing activity of the recombinant strain *E. coli* PWS1 can be turned on by IPTG induction (*29*). Since mercury is toxic to the mercury-resistant cells when its concentration is higher than some threshold value (*28*), the loading of mercury into the bioreactor need to be well controlled for a stable mercury detoxification operation. In addition, a sufficient initial cell concentration to start up the bioreactor is also critical to enable total survival of the cells for efficient mercury detoxification (*26-28*).

It was found that *P. aeruginosa* PU21 (Rip64) is able to detoxify mercury rapidly during its transient growth (*20,26*). The reduction of mercury with growing cells exhibited a maximal rate (approximately 1.11×10^{-6} µg Hg/cell/h) at the substrate concentration of 8 mg Hg/L. The mercury detoxification performance of *P. aeruginosa* PU21 was also correlated to the bacterial growth phases, with cells at lag phase and exponential phase having higher specific detoxification rates. The recombinant strain *E. coli* PWS1 exhibited better specific mercury detoxification activity than *P. aeruginosa* PU21 at batch

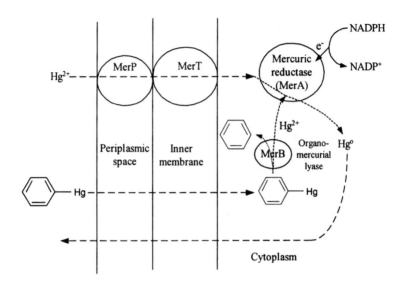

Fig. 1 Mechanism of bacterial mercury resistance

operations, while the PWS1 strain became unstable during long-term fed-batch operations primarily due to plasmid instability (*27*).

Two fed-batch strategies were used for mercury detoxification operations: step-wise fixed-interval (SWFI) feeding (*26*) and repeated constant-rate (RCR) feeding (*27*). For SWFI feeding, mercury detoxification efficiency was 2.9 mg/L/h and 3.3 mg/L/h for 2 and 5 mg Hg/L feeding, respectively. In contrast, for RCR feeding strategy, the mercury detoxification efficiency was closely related to the mercury feeding rate (F_{Hg}), initial inoculum size (X_o), as well as the initial culture volume (V_o) and the best mercury detoxification efficiency was 5.6 mg Hg/L/h, which occurred when F_{Hg} = 16.9 mg Hg/h, X_o = 4.45 x 10^9 cells/ml, and V_o = 1200 ml.

Chemostat operations for mercury detoxification were also conducted but gave lower detoxification efficiencies than those obtained from fed-batch mode (*26,27*). At a dilution rate of 0.18 and 0.325 h^{-1} and mercury feeding concentration of 6.15 mg/L, the mercury detoxification efficiency was only 1.09 and 1.94 mg/L/h, respectively. However, chemostat operations were able to continuously reduce mercury concentrations in the feed (1.0-6.15 mg/L) to nearly undetectable amounts, and the steady-state operation can be maintained for a prolonged period (nearly 55 h).

Mercuric reductase was purified from a recombinant strain *E. coli* PWS1 able to overexpress merA product (*29*). The purified mercuric reductase was covalently immobilized on a porous support Celite 545 and the immobilized enzyme was utilized for mercury detoxification with a fixed-bed process (*30*). Enzymatic reduction of Hg^{2+} with immobilized mercuric reductase requires continuous feeding of cofactor NADPH, which is very costly. Therefore, compounds that have similar structure to NADPH were used as the artificial cofactor. Neutral Red (a dye) was found to be a satisfactory alternative of NADPH as it achieved nearly 40% mercury reduction efficiency compared to using NADPH. The best mercury detoxification performance of the immobilized-enzyme column was 5.41 μmol/h/L (27).

It is of importance to recover the end product of mercury detoxification process, metallic mercury. The end product is mainly in the gas stream and can be recovered by adsorption on metal (e.g., gold, silver, copper) or charcoal adsorbents (7), or by absorption and stabilization with acids. In our recent work (26), mercury vapor was collected by a device consisted of three gas/liquid absorption columns in series using concentrated HNO_3 as the oxidant, followed by an activated carbon adsorption trap. With continuous feeding of up to 10 mg/L of Hg^{2+}, the recovery efficiency in acid columns was over 80% and the mercury concentration at the exit of activated carbon trap was below 0.2 μg/L.

Bioprocess Development for Bacterial Decolorization of Azo Dyes

Azo dyes are one of the most commonly used synthetic dyes in textile, food, papermaking and cosmetic industries (31-33). Release of azo dyes into the environment has become a major concern in wastewater treatment since some azo dyes or their metabolites may be mutagens or carcinogens (34). Azo dyes are often considered xenobiotics and recalcitrant aerobically, and conventional aerobic wastewater treatment processes usually cannot efficiently decolorize azo-dye-contaminated effluents (32,33,35). Therefore, in some cases traditional biological processes are combined with physical and chemical treatment, such as flocculation, chemical coagulation, precipitation, to achieve a better decolorization of the wastewater (36). However, as more efficient bacterial decolorizers have been isolated or constructed, biological decolorization may become an effective method to transform, degrade, or mineralize azo dyes (32,37).

As indicated in Fig. 2, bacterial decolorization of azo dyes is typically initiated by azoreductase-catalyzed anaerobic reduction or cleavage of azo bonds (35), followed by aerobic (38) or anaerobic (39) degradation of the resulting aromatic amines by a mixed bacterial community (40,41). Decolorization by fungi (e.g., *Phanerochaete chrysosporium*) is achieved via the ligninolytic-degradation activity to degrade azo dyes aerobically with the aid of an extracellular enzyme, lignin peroxidase (42). Most biological decolorization processes reported in the literature have been fungi systems (37,43,44), while the development of bacterial decolorization processes has been relatively less studied. Hu (45) recently isolated a *Pseudomonas luteola* strain able to efficiently decolorize a group of azo dyes. The strain was subject to detail investigations in terms of decolorization kinetics and mechanism (46), and decolorization processes including a fed-batch process (47) and an immobilized-cell system (*48*). The author also obtained an *E. coli* mutant (*E. coli* NO3) showing excellent decolorization rate for a variety of azo dyes (*49,50*). In addition, a dye-decolorizing recombinant strain *E. coli* CY1 harboring azo-dye-decolorizing genes of *Rhodococcus* sp. was also constructed in our laboratory (*51*).

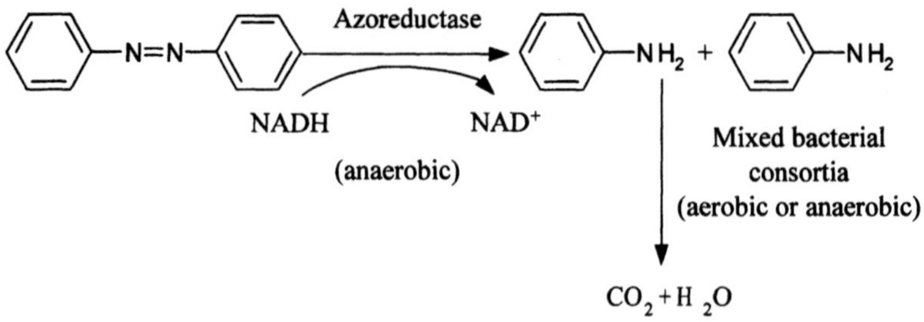

Fig. 2 Mechanism of bacterial decolorization of azo dyes

General Considerations of Bacteria Decolorization Processes

Many aerobic (e.g. *P. luteola*) or facultative anaerobic bacteria (e.g. *E. coli*) show excellent decolorization activity for azo dyes when the dissolved oxygen is limiting, since the presence of free oxygen seems to inhibit azoreductase activity (*32*) due to competing with azo bonds for electron donors (e.g. NADH). Therefore, during the course of decolorization, cell growth essentially stops or becomes very limited as a result of lack of oxygen supply. This appears to cause difficulty in bioreactor design, because maintaining sufficient cell concentration of the bacterial decolorizer in the reactor would be a problem if the cells do not grow at a satisfactory rate under the reaction conditions. As a result, for stable operation of a long-term continuous decolorization bioprocess, it is necessary to develop strategies to provide oxygen for cell growth without hindering decolorization activity or strategies to maintain an adequate amount of cells in the bioreactor when cell growth is insignificant.

On the other hand, pH value is also a critical parameter that affects the efficiency of bacterial decolorization. For instance, a pH value of 7-9 favored decolorization of Reactive Red 22 by *P. luteola* and *E. coli* (*46,48*), while a low pH severely limited the decolorization activity. It was also found that when glucose was used as the carbon source for *P. luteola* and *E. coli* NO3, the pH dropped significantly, probably due to acid formation (e.g., acetic acid), to inhibit their decolorization activity. The inhibitory effect was not observed when an alternative carbon source, such as glycerol, was used (*46*).

Decolorization with Fed-batch Processes Utilizing High-Cell-Density Fermentation

Fed-batch bioprocesses using *P. luteola* as the decolorizer were shown to effectively decolorize an azo dye, Reactive Red 22 (*47*). To avoid oxygen inhibition on the bacterial decolorization, the bioreactor was operated with gentle agitation and without aeration, but the condition led to insignificant cell growth. Initial biomass loading, the feeding strategy, and the yeast extract to dye ratio (Y/D) all affected the performance of the fed-batch bioreactor, with the dye loading rate the most critical one. Since cells grew poorly under oxygen-deficient environment required for decolorization, a sufficient biomass loading is crucial to provide enough total decolorization activity for color removal. High-cell-density fermentation strategies was applied to increase the initial biomass loading to up to 15 g/L and resulted in a two-fold increase in overall decolorization efficiency. Within a certain limit, the decolorization efficiency was enhanced significantly by increasing the dye loading rate, while overloading of the dye may cause substrate inhibition and may also result in higher residual dye concentration at the end of feeding, raising difficulty in the follow-up treatment. The Y/D ratio is also an important factor to ensure sufficient supply of yeast extract, which serves as the raw material for electron-donor (NADH) production. It was found that Y/D ratio should exceed 0.5 to maintain a stable decolorization (*47*). Without using high-cell-density fermentation approaches (initial biomass loading = 2 g/L), the best specific decolorization rate and overall decolorization efficiency obtained for *P. luteola* were 113.7 mg dye/g cell/h and 86.3 mg dye/l//h, respectively, with a constant dye loading rate of 200 mg/h and a Y/D ratio of 1.25. In contract, when initial biomass loading was increased to 15 g/L, the overall efficiency raised to 194 mg dye/g cell/h, while best specific rate was comparable to that obtained without employing high-cell-density fermentation.

Continuous Decolorization with Aerobic/Anaerobic Sequences

Since cell growth is poor during conditions for efficient decolorization, it is not feasible to apply a regular continuous culture for decolorization of azo dyes. A modified process that includes an aerobic growth stage and an anaerobic decolorization stage was examined for its feasibility for persistent and stable decolorization operations. Description of the operation strategy is indicated in Fig. 3. The cells were first grown aerobically in the reactor at batch mode until cell concentration exceeded 2.5 g/L. The reactor was then switched to continuous mode without air supply but with gentle agitation. The continuous culture was operated at different dilution rates to examine the effect of substrate (medium and dye) loading on decolorization. Because the continuous culture was carried out under oxygen-limiting condition, cell concentration in the bioreactor decreased progressively with operation time. When cell concentration was lower than 0.4 g/L, the feeding and effluent were terminated, and the aeration was turned on for aerobic cell growth. Again, when cell concentration reached the designated value, the reactor was returned to continuous mode. This procedure was repeated several times.

The intermittent anaerobic-aerobic process was used to decolorize mixed dye with *E. coli* NO3 strain at hydraulic retention time (HRT) of 32, 18, and 10 h and a constant dye feeding of 5000 ADMI. The color-removal efficiency was around 70 - 80% at all runs for nearly 120 h of operation. The total decolorization efficiency was 93, 156, and 65 ADMI · L/h at HRT=32, 18, and 10 h, respectively. If combination of one aerobic stage and one anaerobic stage is defined as one cycle, the decolorization rate of *E. coli* NO3 appeared to become higher after the original cycle. This may be attributed to the acclimation effect since the *E. coli* NO3 cells were repeatedly exposed to the azo dyes.

Decolorization with Continuous Cultures Combined with Microfiltration Membrane

Since bacterial decolorization normally proceeds under anaerobic conditions, which limit cell growth of the bacterial decolorizers. Therefore, retention of cells within the bioreactor becomes a critical issue in performing a successful continuous decolorization processes. One way to enhance cell retention is to completely recycle of the cells with a microfiltration membrane. Cross-flow microfiltration modules (e.g., hollow fiber membrane) have been frequently used due to its capability for a long-term operation. The membrane reactor causes no mass transfer limitations and also raises the cell concentration within the reactor (even though the growth rate is low); thus has the potential to enhance the operation efficiency.

In our lab, a membrane bioreactor was designed for continuous decolorization of dye-containing water using *P. luteola* as the decolorizer. Figure 4 shows organization and set-up of the bioreactor. Cells were first grown in the reactor at a batch mode until cell density reached a desired level, then the reactor was switched into a continuous mode as the feeding (medium and dye) started and the effluent was introduced into a hollow-fiber membrane module, where cells were blocked and returned into the bioreactor, but the decolorized water penetrated through the membrane. The optimal operation strategy was determined by investigating the effects of operation parameters (e.g., temperature, hydraulic retention time (HRT), dye loading rate, medium loading rate, and biomass concentration, etc.) on decolorization efficiency. The results show that the best decolorization rate occurred at a temperature of 40°C. The overall decolorization efficiency increased when biomass concentration was increased. The optimal HRT was 13 h. The decolorization rate increased rapidly with an increase in dye loading rate. The best specific decolorization rate (90 mg dye/g cell/h) was obtained at a dye loading of 50 mg/h. Addition of extracellular metabolites of a dye-decolorizing bacterium *E. coli* NO3 (*49,50*) appeared to enhance the decolorization ability of *P. luteola*, and the maximal decolorization was elevated from 90 to 125 mg dye/g cell/h.

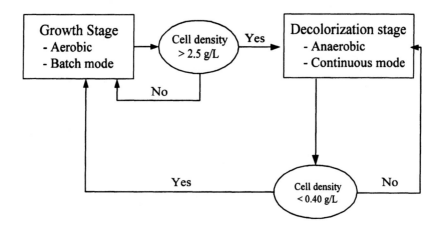

*Fig. 3 Operation strategy for continuous decolorization with aerobic/anaerobic
sequences*

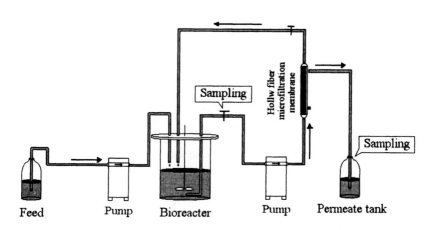

*Fig. 4 Schematic description of membrane bioreactor for continuous
decolorization of azo dye by P. luteola.*

Decolorization with Fixed-Bed Bioreactor containing Immobilized Cells

Cell immobilization is an alternative way to retain cells inside the bioreactor for continuous decolorization. Cell immobilization systems provide an easier solid/liquid separation, which is often encountered in bioprocess operations. Entrapment of cells in natural or synthetic gel matrices allows efficient cell retention and is usually cheaper than the membrane systems. However, the mechanical stability and the effect of the matrix materials or synthesis procedures on cell viability need to be concerned. Entrapment of cells may also raise mass transfer resistance between cells and substrates. Biofilm systems, in which cells attached on the surface of the carrier support, are also one of the most commonly applied cell immobilization approaches, especially in biological treatment of environmental pollutants (*3*). Establishment of a biofilm system does not require extra physical or chemical treatment, making it very friendly to the environment and is also cost-effective. However, biofilm formation is usually time-consuming and the attached cells may come off easily due to shear stress caused by high volumetric flow rates.

Gel Entrapment Systems

Two natural gel matrices (calcium alginate (CA) and κ-carrageen (CN)) and a synthetic polymer matrix (polyacrylamide; PAA) were used to entrap the cells of *P. luteola* (*48*) and the resulting immobilized cells were packed into glass columns for fixed-bed decolorization of an azo dye. The two natural gels are easy to prepare but the beads are less stable mechanically. In addition, the CA beads are sensitive to single-charged cations (e.g. Na^+, K^+), while elevated temperature causes damages to the structure of CN beads. The synthetic gel (PAA) has better mechanical strength but its monomer (arcylamide) is toxic to the cells. The bioprocess was conducted at different dye loading (C_{dye}), bed length (L), representing biomass loading, and volumetric flow rate (F). The CA and PAA columns gave much better decolorization performance than the CN column did. The best decolorization efficiency for CA column at L= 20 cm was 15 mg/h/L (C_{dye}=50 mg/L, F=30 ml/h) and for PAA column was 37 mg/h/L (C_{dye}=200 mg/L, F=15 ml/h).

Biofilm Systems

Decolorization was also conducted with *P. luteola* cells forming biofilm on ceramic supports. Figure 5 shows the SEM micrograph of the support surface before and after biofilm formation. The support materials were packed into a column (5 cm in diameter, 30 cm in length), which was fed with inoculated growth medium for ca. 2 days to form sufficient amount of biofilms. The biofilm-loaded column was used for fixed-bed decolorization operations. The results show that the reactor system maintained 95% decolorization efficiency at low dilution rates (below 0.25 h^{-1}) with an influent dye concentration of 200 mg/L. However, as HRT exceeded 0.25 h^{-1}, decolorization became less efficient,

(a)

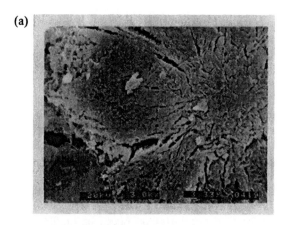

(b)

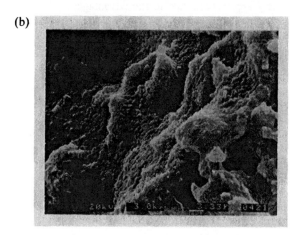

Fig. 5 SEM micrograph of ceramic support surface: (a) before biofilm formation; (b) after biofilm formation.

but addition of the metabolites from *E. coli* NO3 allowed the decolorization efficiency of the biofilm system returned to 95%. The best specific decolorization rate for the operation without metabolite supplements was 2.2 mg/g cell/h (dilution rate = 0.25 h^{-1}), while the rate increased to 5.6 mg/g cell/h (dilution rate = 0.5 h) when 40% (v/v) of metabolites from *E. coli* NO3 were added into the feed.

Conclusions

This chapter demonstrates bioreactor strategies for microbial detoxification of mercury and decolorization of azo dyes. Due to differences in the characteristics of the target substrate and the environment required for the biotransformation, the key concerns for bioreactor design in the two cases are different. For mercury detoxification, fed-batch bioreactors were shown to be favorable due to more efficient control of mercury concentration for optimal detoxification and also to avoid substrate inhibition arising from toxic effects of mercury. The best mercury detoxification efficiency was 5.6 mg Hg/h/L, which occurred when the substrate was fed with a repeated-continuous-rate strategy. For azo-dye decolorization, since cell growth and decolorization take place at distinct oxygenic states, the challenge becomes to determine the timing of oxygen supply for cell growth and to maintain high cell density during decolorization stage, in which cells essentially do not grow. Among the bioreactor strategies examined, the immobilized-cell systems utilizing hollow-fiber membrane and biofilm on porous supports are better bioprocesses for continuous decolorization. Although the membrane reactors are able to maintain high cell density for high decolorization rates, the biofilm systems are more feasible in practical applications due to the advantages of being cost effective and easier to operate and scale up. Moreover, development of better decolorizing biocatalysts by molecular genetic approaches would have the potential to enable further improvement in decolorization efficiency.

References

1. Zitomer, A. H. and Speece, R. E. *Environ. Sci. Technol.* **1993**,. *27*, 227-244.
2. White D. *The physiology and biochemistry of prokaryotes*; Oxford University Press, Inc., New York, NY, 1995.
3. Rittmann, B. E. and McCarty, P. L. *Environmental Biotechnology*; McGraw-Hill Co., Inc., New York, NY, 2001.
4. D'Itri, F. M., Eds. *The Environmental Mercury Problem.* CRC press Cleveland, OH, 1972
5. Noyes, O.R., Hamdy, M. K., and Muse, L. A., *J. Toxicol. Environ. Health* **1976**, *1*, 409-420.
6 Nriagu, J.O., Eds. *Biogeochemistry of mercury in the environment*, Elsevier/North-Holland Biomedical Press, N.Y., 1979.
7. Jones, H.R., Eds. *Mercury Pollution Control*; Noyes Data Corp., Park Ridge, NJ. 1971.
8. Silver, S., Misra, T. K., and Laddaga, R. A. *Bacterial resistance to toxic heavy metals*, pp. 121-139. In: Metal Ions and Bacteria. T. J. Beveridge,e Eds. John Wiley & Sons, New York, NY, 1989.
9. Summers, A.O. and Silver, S. *Ann. Rev. Microbiol.* **1978**, *32*, 637-72.
10. Summers, A.O. and Silver, S., *J. J. Bacteriol.* **1972**, *112*, 1228-1236.
11. Wood, J. M. and Wang, H. K. *Environ. Sci. Technol.* **1983**, *17*, 582A-590A.
12. Foster, T. J. *CRC Crit. Rev Microbiol.* **1987**, *15*, 117-140.

13. O'Halloran, T. V. *Science* **1993**, *261*, 715-725.
14. Summers, A.O. *Ann. Rev. Microbiol.* **1986**, *40*, 607-634.
15. Belliveau, B. H. and Trevors, J. T. *Appl. Organometal. Chem.* **1989**, *3*, 283-294.
16. Misra, T. K. *Plasmid* **1992**, *27*, 4-16.
17. Robinson, J. B. and Tuovinen, O. H. *Microbiol. Rev.* **1984**, *48*, 95-124.
18. Lund, P. A. and Brown, N. L. *Gene* **1987**, *52*, 207-214.
19. Clark, D. L., Weiss, A. A., and Silver, S. *J. Bacteriol.* **1977**, *132*, 186-196.
20. Chang, J.-S. and Hong, J. *J. Biotechnol.* **1995**, *42*, 85-90.
21. Philippidis, G. P., Schottel, J. L., and Hu, W.-S. *Biotechnol. Bioeng.* **1991**, *37*, 47-54.
22. Philippidis, G. P., Schottel, J. L., and Hu, W. S. *Enzyme Microb. Technol.* **1990**, *12*, 854-859.
23. Saouter, E., Turner, R., and Barkay, T. *Annals N.Y. Acad. Sci.* **1994**, *721*, 423-427.
24. Ghosh, S., Sadhukhan, P. C., Chaudhuri, J., Ghosh, D. K., and Mandal, A. *J. Appl. Bacteriol.* **1996**, *81*, 104-108.
25. Ghosh, S., Sadhukhan, P. C., Ghosh, D. K., Chaudhuri, J., and Mandal, A. *Bull. Environ. Contam. Toxicol.* **1996**, *56*, 259-264.
26. Chang, J.-S. and Law, W.-S. *Biotechnol. Bioeng.* **1998**, *57*, 462-470.
27. Chang, J.-S., Chao, Y.-P., and Law, W.-S. *J. Biotechnol.* **1998**, *64*, 219-230.
28. Chen, B.Y. and Chang, J.-S. *Bioprocess Eng.* **2000**, *23*, 675-680.
29. Chang, J.-S. , Chao, Y.-P., Fong, Y.-M., Hwang, Y.-P. and Lin, P.-J. *J. Chin. Inst..Chem. Eng.* **1998**, *29*, 265-274.
30. Chang, J.-S., Hwang, Y.-P., Fong, Y.-M, and Lin, P.-J. J. Chem. Technol. Biotechnol. **1999**,*74*, 965-973.
31. Michaels G. B. and Lewis D. L. *Environ. Toxicol. and Chem.* **1986**, *5*, 161-166.
32. Chung K.-T. and Stevens S. E. Jr. *Environ. Toxicol..Chem.* **1993**, *12*, 2121-2132.
33. Carliell C. M., Barclay S. J., Naidoo N., Buckley C. A., Mulholland D. A. and Senior E. *Water SA* **1995**, *21*, 61-69.
34. Heiss G. S., Gowan B. and Dabbs E. R. *FEMS Microbiol. Lett.* **1992**, *99*, 221-226.
35. Zimmermann T., Kulla H. G. and Leisinger T. *Eur. J. .Biochem.* **1982**, *129*, 197-203.
36. Vandevivere P. C., Bianchi R. and Verstraete W. *J. .Chem. Technol. Biotechnol.* **1998**, *72*, 289-302.
37. Banat I. M., Nigam P., Singh D. and Marchant R. *Bioresource Technol.* **1996**, *58*, 217-227.
38. Seshadri S., Bishop P. L. and Agha A. M. *Waste Management* **1994**, *14*, 127-137.

39. Flores E. R., Luijten M., Donlon B. A., Lettinga G. and Field J. A. *Environ. Sci. Technol.* **1997**, *31*, 2098-2103.
40. Haug W., Schmidt A., Nortemann B., Hempel D. C., Stolz A. and Knackmuss, H.-J. *Appl. Environ. Microbiol.* **1991**, *57*, 3144-3149.
41. Coughlin M. F., Kinkle B. K., Tepper A. and Bishop P. L. *Water Sci. Technol.* **1997**, *36*, 215-220.
42. Glenn J. K. and Gold M. H. *Appl. Environ. Microbiol.* **1983**, *45*, 1741-1747.
43. Yang F. and Yu, J. *Bioprocess Eng.* **1996**, *16*, 9-11.
44. Zhang F.-M, Knapp J. S. and Tapley K. N. *Enzyme Microbial Technol.* **1999**, *24*, 48-53.
45. Hu, T. L. (1994) *Bioresource Technol.* **1994**, *49*, 47-51.
46. Chang, J.-S., Chou, C., Lin, Y.-C., Ho, J.-Y., Lin, P.-J. and Hu, T. L. *Water Res.* **2001**, *35*,.2841-2850
47. Chang, J. S. and Lin, Y.-C. *Biotechnol. Prog.* **2000**, *16*, 979-985.
48. Chang, J.-S., Chou, C., and Chen, S. Y. *Process Biochem.* **2001**, *36*, 757-763.
49. Chang, J. S., Kuo, T.-S. *Bioresource Technol.* **2000**, *75*, 107-111.
50. Chang, J. S., Kuo, T.-S., Chao, Y.-P., Ho, J.-Y., and Lin, P.-J. *Biotechnol. Lett.* **2000**, *22*, 807-812.
51. Chang, J.-S., and Lin, C.-Y. *Biotechnol. Lett.* **2001**, *23*, 631-636.

Chapter 11

Vectors for Genetic Engineering of Corynebacteria

Kaori Nakata, Masayuki Inui, Peter B. Kos, Alain A. Vertès, and Hideaki Yukawa*

Research Institute of Innovative Technology for the Earth, 9–2 Kizugawadai, Kizu, Soraku, Kyoto 619–0292, Japan

Corynebacterium glutamicum and related species are microorganisms of biotechnological significance. These bacteria have a long history of use for the production of fine chemicals such as amino acids and organic acids. Moreover, they have recently been contemplated for the production of xenogeneic proteins, such as proteins from Mycobacteria of medical importance, given the relatively close phylogenetic relationship that exists between these two genera. In this paper, we review various molecular tools available to manipulate Corynebacteria, provide DNA sequence information on several useful vectors, and report on novel shuttle vectors for the genetic engineering of these organisms. Particularly, this series of novel plasmids enables the use of chloramphenicol, kanamycin, gentamycin, or spectinomycin resistance for positive selection in Corynebacteria. These vectors complement the various molecular biology tools available to engineer these important bioconverters and thus pave the way to their increased industrial applications. Moreover, we present evidences corroborating the view that the *B. stationis* plasmid pBY503 belongs to a novel family of rolling-circle plasmids comprising pNG2 isolated from *C. diphtheriae*.

Corynebacterium glutamicum is a gram-positive bacterium with moderately high GC content. It is an organism that has a long history of use in biotechnology for the production of amino acids, which represent for animal feed additives alone a global market of more than \$2 billions (*1-3*). Furthermore, the potential of corynebacteria for xenogeneic protein production appears promising (*4*). The recent sequencing of its complete genome will facilitate the genetic engineering of this important bioconverter, for example by enabling rational engineering (*5*) as a complementary tool to classical whole-cell mutagenesis or genome shuffling (*6*). The numerous strains that have been isolated as glutamate-producing organisms constitute species of *C. glutamicum* (*7*), which, in the *Corynebacterium-Mycobacterium-Nocardia* cluster are closely related to *C. acetoacidophilum*, *C. callunae*, and *C. ammoniagenes*. The clinical species most closely related are *C. minutissimum*, *C. striatum*, and *C. xerosis* (*8, 9*).

Genetic tools have been developed to manipulate the amino acid producing coryneform bacteria. For instance, electroporation protocols that enable the efficient transformation of corynebacterial cells at a high efficiency have been reported by several groups (*10-14*). Such a protocol was also developed in order to engineer *C. glutamicum* R, a strain that exhibits several useful biotechnological characteristics, including the ability to grow on ethanol as a sole carbon source, no autolysis under starvation conditions, and a rapid growth rate. Using the protocol described by Kurusu et al. (*15*), *C. glutamicum* R cells were successfully transformed to an efficiency of approximately 10^4 transformants per μg of DNA. Typically, in this procedure, cells in mid-log phase were prepared for electroporation by incubation with penicillin G as a means to weaken the cell wall. A major barrier to electroporation was demonstrated to be the restriction barrier, which is due, in *C. glutamicum* MJ233C, to an *mrr*-like restriction system, and to an *mcr*-like system in *C. glutamicum* ATCC 31831. On the other hand, no such system could be identified in *C. glutamicum* ATCC 13869 (*16*). When plasmid DNA was isolated from coryneform cells, or from a *dam dcm E. coli* mutant such as strain JM110 (*17*), transformation efficiencies up to 4.0×10^7 transformants per μg DNA were typically attained (*16*).

Conjugation is an alternative protocol that has been developed by several researchers (*18-21*) as conjugative plasmid DNA transfer has been observed between Gram-positive and Gram-negative bacteria (*22*). The success of conjugation as a gene transfer method is also explained by the observation that it enables to bypass, to some extent, the restriction barrier. Schäfer et al. (*20*) and van der Rest et al. (*14*) further used stress exposure, a well-described method to temporarily inactivate the restriction barrier, as a means to increase conjugation frequencies.

The advent of these efficient transformation techniques enabled the use of integrative plasmid DNA in order to perform gene disruption and replacement in this group of important amino acid producers (*23*). Many of these protocols rely on conjugative transfer of plasmids only bearing an *E. coli* origin of replication. An electroporation-based protocol was developed for *C. glutamicum* MJ233C that is also suitable for the manipulation of other strains of coryneform bacteria, reaching transformation efficiencies with integrative plasmids of up to 2.4×10^2 transformants per μg DNA (*24*).

In recent years, several useful cloning vectors, designed around endogenous plasmids (*25*), have been constructed, including gene expression shuttle vectors (*21, 26-29*). We previously reported on pBY503, a 16.4-kb cryptic plasmid from *Brevibacterium stationis* (*30*) and associated partition function (*31*), plasmid pPROBE17, a promoter probe vector (*32*), pMV5, a transposon trap vector (*33*) based on the lethal phenotype conferred by the *B. subtilis sacB* gene (*34*), and pMV11 (Tn*31831*) and pMV23 (miniTn*31831*), two mobile elements delivery vehicles (*35*).

We report in this paper on the construction of novel shuttle vectors for the manipulation of *C. glutamicum*.

Materials and Methods

Bacterial Strains and Culture Conditions

E. coli JM109 (*17*), *E. coli* JM110 (*17*), and *C. glutamicum* R, a plasmidless strain from the laboratory collection, were used in this study. Plasmids pBL1 and pMV5 were from the laboratory collection. Plasmids pHSG299 and pHSG398 were obtained from Takara (Osaka, Japan). *E. coli* strains were grown at 37°C in LB medium supplemented with 50 μg.ml^{-1} ampicillin, 50 μg.ml^{-1} chloramphenicol, 50 μg ml^{-1} kanamycin, 10 μg ml^{-1} gentamycin, or 200 μg ml^{-1} spectinomycin as appropriate. Corynebacteria were routinely grown at 33°C in AR Medium (*15*). Kanamycin, chloramphenicol, gentamycin, and spectinomycin were added as appropriate (50 μg ml^{-1}, 5 μg ml^{-1}, 10 μg ml^{-1}, and 200 μg ml^{-1}, respectively).

DNA techniques

Chromosomal DNA was isolated using the Amersham Pharmacia Biotech (UK) Genomic Prep kit. Plasmid DNA was isolated by the alkaline lysis procedure (*36*). Restriction endonucleases, Klenow fragment, and T4 DNA ligase were obtained from Takara (Osaka, Japan) and used as per the manufacturer's instructions. Restriction fragments were isolated when required from agarose gels with the Prep-a-Gene matrix (Bio-Rad, Richmond, CA) according to the manufacturer's instructions. All PCR were carried out in an Applied Biosystems GenAmp System 9700 as recommended by the manufacturer using the following amplification protocol: 30 cycles at temperatures of 95°C for denaturation (1 min), 54°C for annealing (1 min), and 72°C for extension (1 min).

Bacterial cells transformation

Corynebacteria were transformed by electroporation (*16*) using plasmid DNA purified from *E. coli* JM110. *E. coli* strains were transformed by the $CaCl_2$ method (*36*).

DNA sequencing

We have generated a library of the DNA fragments to be sequenced using a Misonics sonicator (Farmingdale, NY) on power setting 1. The Eppendorf tube containing the DNA sample was placed in an ice bath and 8 cycles of sonication for 1 sec interrupted by 1 sec intervals were performed. The resulting random fragments were size separated on a 1% agarose gel and the fraction corresponding to the 2-3 kb pool was subsequently purified from the gel. The fragments were blunted by treatment with the Klenow fragment and ligated to *Sma*I digested pUC119 plasmid DNA (obtained from Takara). The ligation mixture was used to transform *E. coli* JM109, recombinants were selected on IPTG-supplemented plates (*36*). For sequencing purposes, clones were grown overnight in 96 deep-well microtiter plates in 1 ml 2X-LB medium using a TAITEC Bioshaker. The corresponding plasmids were isolated in 96 well plates using the Millipore Montage Plasmid Miniprep96 kit and a Cosmotec HT Station 500 (Tokyo, Japan) simple 96-well format pipetting robot. The plasmid library thus generated was sequenced on both ends using M13 universal forward and reverse primers (*36*) and cycle sequencing using the BigDye method of ABI Biosystem Inc. in an ABI 3700 CE sequencer. The Applied Biosystems GeneAmp System 9700 was used for polymerase chain reactions as described in the DNA Techniques section. Prior to sequencing, the resulting PCR products were treated with exonuclease using the ExoSAP-IT from USB (Cleveland, OH, USA) to destroy remaining primers.

DNA sequences were analyzed as follows. The raw chromatogram files (.abi files) were collected on a personal computer running Windows 2000 (Microsoft Corp.). The chromatogram files were subsequently transferred to a SunBlade 1000 computer (Sun) with one 900 MHz 64-bit UltraSparc-III CPU, 2 GByte memory. The PREGAP4 program of the Staden package (*37, 38*) was used for clipping vector sequences, as well as for quality clipping and contamination screening after base-calling by PHRED (*39, 40*). BLASTX searches were carried out on the Sun computer using the stand-alone BLAST program (*41*) of NCBI (http://www.ncbi.nlm.nih.gov) using matrix BLOSUM62 (*42*).

Nucleotide accession number

The nucleotide sequences reported in this paper appears in the EMBL and GenBank Nucleotide Sequence Databases under the accession numbers AY211882 (pCRA1) and AY211883 (pMV5).

Results

Construction of the pCRA1 *E. coli* – *C. glutamicum* shuttle vector

Plasmid pCRA1 is an *E. coli* – *C. glutamicum* shuttle vector that derives from the *E. coli* plasmid pHSG398 (Takara) and the coryneform plasmid pBL1 (*43*). Plasmid pCRA1 is 5.3-kb in size and confers chloramphenicol resistance to both organisms. Plasmid pBL1 was independently isolated by several group and reported also as pAM330 (*44*), pWS101 (*45*), and pGX1901 (*46*). It is a multicopy plasmid 4.46-kb in size that replicates via a rolling-circle mode (*47*). Plasmid pCRA1 was constructed as follows. Plasmid pBL1 linearized with *Hind*III was ligated to *Hind*III linearized pHSG398 (Takara) plasmid DNA resulting in plasmid pC1. Plasmid pC1 contains two open reading frames derived from pBL1 that are not necessary for replication in *C. glutamicum*, but cause filamentation in *E. coli* (*48*). A 1.4-kb fragment containing these two ORFs was thus excised with *Hpa*I and *Pst*I. Following treatment with the Klenow fragment to blunt the cohesive ends, the plasmid was recircularized using T4 DNA ligase, resulting in the shuttle vector pTK1. A linker comprising the sites *Kpn*I-*Xho*I-*Bgl*II-*Pst*I-*Sma*I(*Xma*I)-*Bam*HI was generated by annealing the following oligonucleotides: CTC GAG ATC TGC AGC CCG G and <u>GAT C</u>CC GGG CTG CAG ATC TCG AG<u>G TAC</u>. The half *Kpn*I and *Bam*HI sites are indicated in underlined characters. Annealing was performed by incubating the oligonucleotides at 75°C for 5 min and at room temperature until equilibrium. The linker was subsequently ligated to *Kpn*I and *Bam*HI digested plasmid DNA resulting in the *E. coli* – *Corynebacterium* shuttle vector pCRA1. Plasmid pCRA1 was sequenced as described in the Materials and Methods section. A restriction and genetic map of this novel shuttle vector is given in Fig. 1. The DNA sequence of the polylinker and flanking sequences is shown in Fig. 2.

Plasmid pCRA1 is an ideal vector for high-level expression; nevertheless, due to its mode of replication, segregational and structural instability may be observed (*49*, *50*), particularly when overexpressing genes whose activities are deleterious to the host strain. For cloning such genes, plasmids that replicate via a *θ* mechanism provide a clear advantage as these plasmids typically benefit from higher structural and segregational stability (*50*).

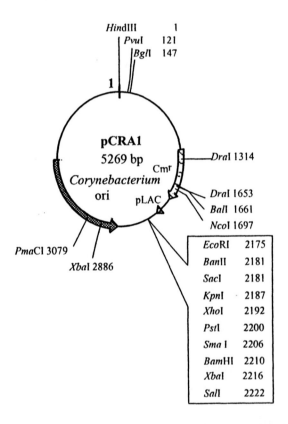

Figure 1. The physical and genetic map of the E. coli – C. glutamicum shuttle vector pCRA1. Only the sites that are either unique or present in two copies are indicated. Arrows give the direction of transcription. The lac promoter is represented by pLAC, the replication origin encoding the protein coding gene by ori, and the chloramphenical resistance gene by Cm'. MCS: multiple cloning site, the order of the restriction sites is given clockwise in the box. Numbers given for each site are coordinates.

```
          10 SqaI  20        30        40        50        60        70        80        90       100       110       120
GAATTACAACAGTACTGCGATGAGTGGCAGGGGCGGGGCGTAATTTTTTAAGGCAGTTATTGGTGCCCTTAAACGCCTGGTCGTACGGCTGAATAAGTGATAATAAGCGGATGAATGGCA

         130       140       150       160       170       180       190       200       210       220       230       240
GAAATTCAGCTTGGCCCAGTGCCAAGCTCCAATACGCAAACCGCCTCTCCCCGCCGGTTGGCCGATTCATTAATGCAGCTGGCACAGAGGTTTCCCGACTGGAAAGCGGGGCAGTGAGCG

         250       260       270       280       290       300       310       320       330       340       350       360
CAACGCAATTAATGTGAGTTAGCTCACTCATTAGGCACCCCAGGCTTTACACTTTATGCTTCCGGCTCGTATGTTGTGTGGAATTGTGAGCGGATAACAATTTCACACAGGAAAACATTG

         370       380       390       400       410       420       430       440       450       460       470       480
ACCATGATTACGCCGAATTCGAGCTCGGTACCTCGAGATCTGCAGCCCGGGATCCTCTAGAGTCGACCAACGTCAACAACCACCCCGCAGCGTTAAGTTGCCCGCCAACAGAAAGGAA
            EcoRI SacI  KpnI XhoI BglII PstI  SmaI BamHI   SalI  PshAI

         490       500       510       520       530       540       550       560       570       580       590       600
ACCAACACGAAACAACAACCAAAAAGTTTCACAGAAAAAAGGCGTATGCGGTAACGTATGCCGTAACGCGGTTAAGCCCTAGCCCGCCGCGCGTAGGTATTACTCA

         610       620       630       640       650       660       670       680       690       700       710       720
TGCCCACTATGGTGTGCACACTGCCCACTACGGTGTGCAATCTATTCACGATGCCACCCCCAGATACAGTGAAGCCCGCCAATCCGAACTAGATCAACGGGCAACCCATTGTC

         730       740       750
CCCAGCTTTGATTAGGAGCCAGGCACATAA
```

Figure 2. DNA sequence of the polylinker of pCRA1 and its flanking regions. The lac promoter is indicated in bold letters. Cm^r: chloramphenicol resistance, Km^r: kanamycin resistance, Gm^r: gentamycin resistance, Spec^r: spectinomycin resistance. The various restriction sites in the polylinkers are given clockwise in the boxes adjacent to each plasmid together with the coordinates.

The pCRB shuttle *E. coli* – *Corynebacterium* shuttle vector series

The design used to construct vector pCRA1 was optimized in order to generate a series of versatile shuttle vectors enabling rapid cloning using *lacZα* complementation in *E. coli* and a variety of positively selectable markers (chloramphenicol, kanamycin, gentamycin, or spectinomycin). The restriction maps and genetic maps of these vectors, pCRB1, pCRB2, pCRB3, and pCRB4, are presented in Fig. 3.

Plasmid pCRB1 was constructed as follows. Primers 2 and 3 (respectively CTC T<u>GA</u> <u>TAT</u> <u>C</u>GT TCC ACT GAG CGT CAG ACC, CTC T<u>GA</u> <u>TAT</u> <u>C</u>TC CGT CGA ACG GAA GAT CAC; the bases corresponding to the restriction sites used in cloning have been underlined) were used to amplify by PCR the entire pHSG398 plasmid. The resulting amplicon was digested with *Eco*RV and ligated to *Hpa*I linearized pBL1 plasmid DNA. The ligation mixture was used to transform *E. coli* and one of the resulting clones was selected randomly as a source of pCRB1 plasmid DNA, which is 4,050-bp in size.

Plasmid pCRB2 was constructed by linearization of plasmid pHSG299 DNA using *Stu*I and ligation to a *Hpa*I digested PCR fragment encoding the replication origin of plasmid pBL1. This fragment was generated using the oligonucleotides CTC T<u>GT</u> <u>TAA</u> <u>C</u>ACA TGC AGT CAT GTC GTG CT and CTC T<u>GT</u> <u>TAA</u> <u>C</u>AC AAC AAG ACC CAT CAT AGT. The resulting plasmid is 4,494-bp in size.

The gentamicin resistance gene (*51*) was available on an *Eco*RI cartridge cloned in plasmid pGP704Gm (*52*). Plasmid pCRB1 DNA was used as a template to amplify by PCR an amplicon devoid of the chloramphenicol resistance marker, using primers 1 and 5 (respectively CTC T<u>GA</u> <u>TAT</u> <u>C</u>CA ATA CGC AAA CCG CCT CTC and CTC T<u>GT</u> <u>TAA</u> <u>C</u>AC AAC AAG ACC CAT CAT AGT). The resulting PCR product was digested with *Eco*RV and *Hpa*I and ligated to *Dra*I linearized pGP704Gm plasmid DNA to yield plasmid pCRB3, 4,729-bp in size, recovered from an *E. coli* colony obtained following transformation of HB101 *E. coli* cells.

The 4,155-bp spectinomycin resistance shuttle vector pCRB4 was constructed by ligating the *Xba*I-*Nde*I spectinomycin resistance cassette available on plasmid pMV5 (*33*) to a pCRB1 amplicon devoid of the chloramphenicol marker and obtained as described above. The *Xba*I-*Nde*I fragment was blunted by treatment with the Klenow fragment and ligated to the above described PCR product digested with *Eco*RV-*Hpa*I.

The complete sequences of these four novel vectors have been assembled *in silico* based on sequence information available in public databases.

The transposon trap vector pMV5

Expression of the *B. subtilis* levan sucrase (*53, 54*) has been observed to be lethal to coryneform bacteria when these cells are grown on minimum medium supplemented with 10% sucrose (*33, 34*).

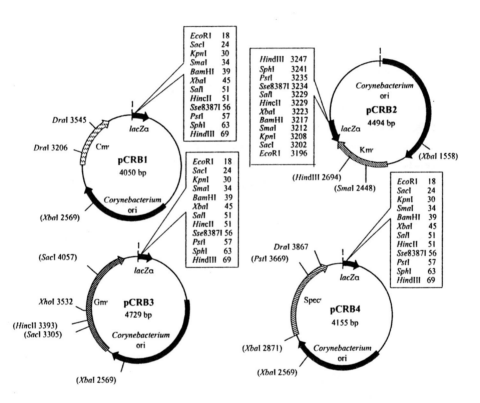

Figure 3. The physical and genetic maps of the E. coli – C. glutamicum shuttle vectors pCRB1, pCRB2, pCRB3, and pCRB4. Only the sites that are either unique or present in two copies are indicated. Arrows give the direction of transcription. Numbers given for each site are coordinates.

Plasmid pMV5 has been constructed to enable the positive selection of replacement recombination events in corynebacteria. Additionally, this vector may prove useful for the cloning of mobile genetic elements from closely related species such as *Rhodococcus* or *Mycobacterium*. Construction of the transposon trap pMV5 has been reported elsewhere (*33*). This vector is based on the replication origin of plasmid pBY503, a 16.2-kb low copy number cryptic episome originating from *Brevibacterium stationis* IFO 12144 (31, 44).

The nucleotide sequence of the replication origin of plasmid pBY503 (GenBank accession number AY211883) was determined using the corresponding fragment borne by plasmid pMV5. This fragment encodes a 1,488-bp long open reading frame whose product shares 79% similarity and 65% identity with the RepA protein of the *C. diphtheriae* plasmids pSV5 (Genbank accession number X57320), pNGA2 (AY061891), and pNG2 (AF492560), 73% similarity and 61% identity with the *C. glutamicum* plasmid pTET3 (*55*), 73% similarity and 60% identity with the *C. efficiens* plasmid pCE2 (AP005225), and 66% similarity and 51% identity with the *C. jeikeium* plasmid pB85766 (AF486522). While the size of plasmid pBY503 and its low copy number of 4.6 per chromosome equivalent in *C. glutamicum* MJ233C (*31*) are characteristics that are typical of plasmids that replicate via a θ mechanism (*50*), these observed homologies suggest that pBY503 belongs to a family of plasmids comprising members that have been demonstrated to replicate via the rolling circle mode (*56*). This plasmid family thus represented by pGA1 (*57*), pSR1 (*58*), pNG2, pNGA2, pSV5, pCE2, pB85766 and pBY503 appears distinct from the family of rolling-circle plasmids originating from staphylococci, streptococci, or lactococci (*59*). The view that pBY503 does not replicate via a θ mechanism is corroborated by the observation that no typical iterons (*60*) could be found upstream of the *repA* gene. In addition, a conserved motif typical for Rep proteins encoded by rolling-circle plasmids and viruses (*61*), uxxYuxKxxx (where u represents a bulky hydrophobic residue such as I, L, V, M, F, Y, or W), is observed in the RepA protein of pBY503 (Table 1), with the notable exception that residue K is replaced by Q. The observation that in addition to pBY503, similar sequences are present in pTET3, pNG2, pNGA2, and pSV5 suggests that Q represents a canonical residue at this particular position. Likewise, it is noteworthy that this plasmid family seems characterized by the sequence VRGYV. Analysis of the sequence furthermore reveals an incomplete additional open reading frame, OrfX, located downstream of *repA* whose gene product exhibits a carboxy terminus that shows strong homologies to transport proteins: 80% similarity to a conserved hypothetical protein of *C. efficiens* YS-314, 78% to a putative permease of *C. glutamicum* ATCC 13032, and 71% to a putative transmembrane transport protein of *Streptomyces coelicolor* A3(2).

Discussion

The genetic engineering of coryneform bacteria, and particularly of *C. glutamicum* R, has been facilitated by recent developments in molecular biology tools and in knowledge of the DNA sequence of its

Table 1. RepA signature motif of rolling-circle plasmids

Plasmids	Host	uxxYuxKxxx motif
pBY503	*B. stationis*	**VRGYVTQSKT**
pTET3	*C. glutamicum*	**VRGYVTQSKT**
pGA1	*C. glutamicum*	**VRGYVAKGQP**
pSR1	*C. glutamicum*	**VRGYVAKGQP**
pEP2 (pNG2)	*C. diphtheriae*	**VRGYVAQSKS**
pNGA2	*C. diphtheriae*	**VRGYVAQSKS**
pSV5	*C. diphtheriae*	**VRGYVAQSKS**
pCE2	*C. efficiens*	**VRGYVTTSKG**
pB85766	*C. jeikeium*	**VRGYVLGNKR**

chromosome (Yukawa et al., unpublished). We describe in this report the construction of versatile *E. coli* –*C. glutamicum* shuttle vectors based on the replication origin of plasmid pBL1, a high copy number episome isolated from various corynebacterial strains and whose replication mechanism involves rolling circles (*47, 48*). These vectors allow *lacZα* complementation in *E. coli*. Furthermore, they bear a multiple cloning site in this latter gene in order to facilitate cloning. The four different resistance markers that characterize this novel shuttle vector series, namely chloramphenicol, kanamycin, spectinomycin, and gentamycin, provide versatile molecular biology tools. These vectors complement the set of existing molecular biology tools for the genetic manipulation of these bacteria, tools that include promoters of a wide range of strengths and specificities, as well as transposon mutagenesis systems. In addition, we present evidences that the *B. stationis* plasmid pBY503 (*30*) belongs to the distinct family of plasmids comprising pGA1 (*57*), pSR1 (*58*), pNG2 (GenBank accession number AF492560), pNGA2 (AY061891), pSV5 (X57320), pCE2 (AP005225), and pB85766 (AF486522). It is noteworthy that the pNG2 miniderivative pEP2 has been demonstrated to replicate via a rolling-circle mechanism (56). Though available observations strongly promote the view that all these plasmids replicate in a similar manner, it would be worth identifying the single strand and double strand origins (50, 62) of these other plasmids using deletion derivatives, and demonstrating that they also replicate via a mechanism similar to that of pNG2. Interestingly, the partition function present in pBY503 is a *cis*-acting element that does not act by increasing plasmid copy number (31), in contrast to the *trans*-acting element recently identified in pGA1, which promotes copy numbers as high as 35 copies per chromosome equivalent (57).

The engineering of corynebacteria is of biotechnological significance for the production of a variety of compounds. For example, our group has successfully produced on an industrial scale numerous amino acids by fermentation, including for instance L-aspartic acid (63), L-isoleucine (64, 65), and L-valine (65). We previously described a novel industrial process for the production of these amino acids. This process is characterized by the use of intact *C. glutamicum* cells under conditions of repressed cell division, and is based on a combination of cell reuse and product recovery as a means to optimize the ratio of substrates transformed into useful products by circumventing the sink of energy into cellular house-keeping cycles and biomass production. In our hands, cells could be reused up to a minimum of 30 times (Fig. 4). This scheme is made possible mainly by the intrinsic property of *C. glutamicum* R that does not undergo autolysis under starvation conditions. Moreover, a variety of methods can be applied to limit by-product formation, such as prior heat treatment of cells in order to inactivate undesirable enzymatic reactions (63), or addition of particular reagents (64).

This process is perhaps best exemplified by the scheme we used for the production of L-aspartic acid (63). *C. glutamicum* (*Brevibacterium flavum*) strain MJ233 is a natural isolate that exhibits a strong aspartase activity. Taking advantage of this property, stoichiometric conversion of fumarate to aspartate was attained by thermal inactivation of fumarase, the enzyme at the origin of by-product (malic acid) formation. Heat treatments of 5 to 9 h at 45°C or of 2 h at 50°C were found to be optimal for achieving a 90% decrease in fumarase activity. Furthermore, calcium ions were the sole divalent metal ions able to activate the aspartase from *C. glutamicum* MJ233, and addition of 0.08% weight/ volume of the detergent Tween 20 (poly-oxyethylene sorbitan monolaurate) was observed to increase aspartase activity up to 40%, presumably via an increase in cell permeability. Likewise, L-aspartic acid at a minimum concentration of 0.6 M was found to protect aspartase during heat treatment. Based on this empirical knowledge and on economic or operational considerations, the industrial aspartic acid production process was designed as a multi-phase process. The first phase comprises a cultivation step to generate biomass and a pre-reaction step to custom-design the intracellular enzyme pool. The biomass generation step is conducted under aerobic conditions at 33°C for 30 h in a pH 7.6 medium consisting of 2.3% $(NH_4)_2SO_4$, 0.05% K_2HPO_4, 0.05% KH_2PO_4, 0.05% $MgSO_4.7H_2O$, 0.3% yeast extract, 0.3% casamino acids, 200 ppb biotin, 100 ppb

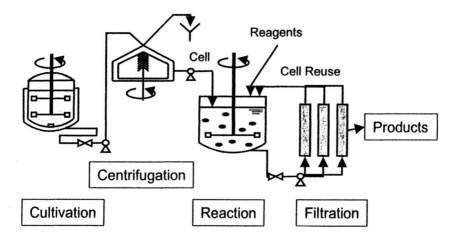

Figure 4. The Living-Cells-Reaction process. This process is characterized by the use of a bacterial production strain that does not undergo autolysis under non-growing conditions. This property renders immobilization unnecessary and enables the use of standard fermentation equipment. Moreover, the rate of carbon utilization towards product production is maximized as vegetative growth and housekeeping functions are minimized. Product separation from the reaction mixture and cell recycling is performed using an ultrafiltration apparatus. Purification of secreted products such as amino acids is facilitated by the negligible contamination from intracellular materials. Moreover, temperatures and pH sub-optimal for growth can be used as a means to limit microbial contamination.

thiamine hydrochloride, 0.002% $FeSO_4.7H_2O$, 0.002% $MnSO_4.nH_2O$, 2% ethanol. The pre-reaction step consists of a heat treatment at 45°C for 5 h in a mixture consisting of 3% weight/ volume cells, 750 mM L-aspartic acid, 2 M ammonium chloride, 7.5 mM calcium chloride, and 0.08% weight/ volume Tween 20. This step is designed to decrease to a negligible level the fumarase activity without significantly affecting the aspartase activity. The second phase is initiated by collecting by centrifugation heat-treated cells and by inoculating with those cells a pH 9.3 reaction mixture consisting of 860 mM fumaric acid, 4 M NH_4OH, 7.5 mM $CaCl_2$. Contamination risks are kept to a minimum by conducting operations at a relatively high temperature (45°C) and at an alkaline pH. The third phase is characterized by cell recycling where, taking advantage of the physiological characteristics of *C. glutamicum* MJ233 (no autolysis and no leakage of intracellular macromolecules), the reaction mixture is separated from the cells by ultrafiltration, typically using a polysulfon hollow-fiber tubular type membrane that can withstand high temperatures and an alkaline pH. As a result, product concentration can reach levels up to approximately 1.3 M. The ability to reach such high product concentrations is an important industrial characteristic of this process as it facilitates the last phase, product purification.

This process thus combines rational and empirical design, and offers several advantages including an easy implementation in standard fermentation equipment, the use of living cells that do not require immobilization, in addition to decreased contamination risks and efficient downstream processing. We coined the word LCR (Living-Cell-Reaction) to encompass all modifications to this basic process, and particularly those industrial schemes where the reaction phase is performed in minimum medium devoid of biotin, an essential growth factor, such that cell growth is halted. This industrial design can be adapted to the production of numerous other fine chemicals such as amino acids or organic acids (64, 65).

Notwithstanding these developments, the industrial use of coryneform bacteria would benefit from the availability of a series of vectors based on a broad-host-range plasmid replicating via a θ mechanism, as, in contrast to typical rolling-circle plasmids (*50, 59*), the intrinsic properties of these episomes include high structural and segregational stability (*60*).

Acknowledgements

This work was supported by a grant from the Ministry of Economy, Trade & Industry (METI).

References

(*1*)   Eggeling, L.; Sahm, H. In *Metabolic Engineering*; Lee, S. Y., Papoutsakis, E. T., Eds.; Marcel Dekker Inc: New York, 1999; pp 153-176.

(2)   Lessard, P. A.; Guillouet, S.; Willis, L. B.; Sinskey, A. J. In *The encyclopedia of bioprocess technology: fermentation, biocatalysis, and bioseparation*; Flickinger, M. C., Drew, S. W., Eds.; John Wiley & Sons: New York, 1999; Vol. 2, pp 729-740.

(3)   Kumagai, H. *Adv Biochem Eng Biotechnol* 2000, *69*, 71-85.

(4)   Billman-Jacobe, H.; Wang, L.; Kortt, A.; Stewart, D.; Radford, A. *Appl Environ Microbiol* 1995, *61*, 1610-1613.

(5)   Ohnishi, J.; Mitsuhashi, S.; Hayashi, M.; Ando, S.; Yokoi, H.; Ochiai, K.; Ikeda, M. *Appl Microbiol Biotechnol* 2002, *58*, 217-223.

(6)   Zhang, Y. X.; Perry, K.; Vinci, V. A.; Powell, K.; Stemmer, W. P.; del Cardayre, S. B. *Nature* 2002, *415*, 644-646.

(7)   Liebl, W.; Ehrmann, M.; Ludwig, W.; Schleifer, K. H. *Int J Syst Bacteriol* 1991, *41*, 255-260.

(8)   Funke, G.; von Graevenitz, A.; Clarridge, J. E., 3rd; Bernard, K. A. *Clin Microbiol Rev* 1997, *10*, 125-159.

(9)   Pascual, C.; Lawson, P. A.; Farrow, J. A.; Gimenez, M. N.; Collins, M. D. *Int J Syst Bacteriol* 1995, *45*, 724-728.

(10)   Bonamy, C.; Guyonvarch, A.; Reyes, O.; David, F.; Leblon, G. *FEMS Microbiol Lett* 1990, *54*, 263-269.

(11)   Bonnassie, S.; Burini, J. F.; Oreglia, J.; Trautwetter, A.; Patte, J. C.; Sicard, A. M. *J Gen Microbiol* 1990, *136*, 2107-2112.

(12)   Dunican, L. K.; Shivnan, E. *Bio/Techniques* 1990, *7*, 1067.

(13)   Haynes, J. A.; Britz, M. L. *J Gen Microbiol* 1990, *136*, 255-263.

(14)   van der Rest, M. E.; Lange, C.; Molenaar, D. *Appl Microbiol Biotechnol* 1999, *52*, 541-545.

(15)   Kurusu, Y.; Kainuma, M.; Inui, M.; Satoh, Y.; Yukawa, H. *Agric Biol Chem* 1990, *54*, 443-447.

(16)   Vertès, A. A.; Inui, M.; Kobayashi, M.; Kurusu, Y.; Yukawa, H. *Res Microbiol* 1993, *144*, 181-185.

(17)   Yanisch-Perron, C.; Vieira, J.; Messing, J. *Gene* 1985, *33*, 103-119.

(18)   Reyes, O.; Guyonvarch, A.; Bonamy, C.; Salti, V.; David, F.; Leblon, G. *Gene* 1991, *107*, 61-68.

(19)   Labarre, J.; Reyes, O.; Guyonvarch, A.; Leblon, G. *J Bacteriol* 1993, *175*, 1001-1007.

(20)   Schafer, A.; Kalinowski, J.; Puhler, A. *Appl Environ Microbiol* 1994, *60*, 756-759.

(21)   Schafer, A.; Tauch, A.; Jager, W.; Kalinowski, J.; Thierbach, G.; Puhler, A. *Gene* 1994, *145*, 69-73.

(22)   Doucet-Populaire, F.; Trieu-Cuot, P.; Andremont, A.; Courvalin, P. *Antimicrob Agents Chemother* 1992, *36*, 502-504.

(23)   Schwarzer, A.; Puhler, A. *Biotechnology (N Y)* 1991, *9*, 84-87.

(24)   Vertès, A. A.; Hatakeyama, K.; Inui, M.; Kobayashi, M.; Kurusu, Y.; Yukawa, H. *Biosci Biotechn Biochem* 1993, *57*, 2036-2038.

(25) Deb, J. K.; Nath, N. *FEMS Microbiol Lett* **1999**, *175*, 11-20.

(26) Eikmanns, B. J.; Kleinertz, E.; Liebl, W.; Sahm, H. *Gene* **1991**, *102*, 93-98.

(27) Martin, J. F.; Santamaria, R.; Sandoval, H.; del Real, G.; Mateos, L. M.; Gil, J. A.; Aguilar, A. *Bio/Technol* **1987**, *5*, 137-146.

(28) Martin, J. F.; Gil, J. A. In *Manual of Industrial Microbiology and Biotechnology*, 2 ed.; Demain, A. L., Davies, J. E., Atlas, R. M., Cohen, G., Hershberger, C. L., Hu, W. S., Sherman, D. H., Willson, R. C., Wu, J. H. D., Eds.; American Society for Microbiology Press: Washington D.C., 1999, pp 379-391.

(29) Takagi, H.; Morinaga, Y.; Miwa, K.; Nakamori, S.; Sano, K. *Agric Biol Chem* **1986**, *50*, 2597-2603.

(30) Satoh, Y.; Hatakeyama, K.; Kohama, K.; Kobayashi, M.; Kurusu, Y.; Yukawa, H. *J Ind Microbiol* **1990**, *5*, 159-165.

(31) Kurusu, Y.; Satoh, Y.; Inui, M.; Kohama, K.; Kobayashi, M.; Terasawa, M.; Yukawa, H. *Appl Environ Microbiol* **1991**, *57*, 759-764.

(32) Zupancic, T. J.; Kittle, J. D.; Baker, B. D.; Miller, C. J.; Palmer, D. T.; Asai, Y.; Inui, M.; Vertès, A.; Kobayashi, M.; Kurusu, Y.; et al. *FEMS Microbiol Lett* **1995**, *131*, 121-126.

(33) Vertès, A. A.; Inui, M.; Kobayashi, M.; Kurusu, Y.; Yukawa, H. *Mol Microbiol* **1994**, *11*, 739-746.

(34) Jager, W.; Schafer, A.; Puhler, A.; Labes, G.; Wohlleben, W. *J Bacteriol* **1992**, *174*, 5462-5465.

(35) Vertès, A. A.; Asai, Y.; Inui, M.; Kobayashi, M.; Kurusu, Y.; Yukawa, H. *Mol Gen Genet* **1994**, *245*, 397-405.

(36) Sambrook, J.; Russell, D. W. *Molecular cloning: a laboratory manual*, 3rd ed.; Cold Spring Harbor Laboratory Press: Cold Spring Harbor, N.Y., 2001.

(37) Bonfield, J. K.; Smith, K.; Staden, R. *Nucleic Acids Res* **1995**, *23*, 4992-4999.

(38) Staden, R. *Mol Biotechnol* **1996**, *5*, 233-241.

(39) Ewing, B.; Hillier, L.; Wendl, M. C.; Green, P. *Genome Res* **1998**, *8*, 175-185.

(40) Ewing, B.; Green, P. *Genome Res* **1998**, *8*, 186-194.

(41) Altschul, S. F.; Madden, T. L.; Schaffer, A. A.; Zhang, J.; Zhang, Z.; Miller, W.; Lipman, D. J. *Nucleic Acids Res* **1997**, *25*, 3389-3402.

(42) Henikoff, S.; Henikoff, J. G. *Proc Natl Acad Sci U S A* **1992**, *89*, 10915-10919.

(43) Santamaria, R.; Gil, J. A.; Mesas, J. M.; Martin, J. F. *J Gen Microbiol* **1984**, *130*, 2237-2246.

(44) Miwa, K.; Matsui, H.; Terabe, M.; Nakamori, S.; Sano, K.; Momose, H. *Agric Biol Chem* **1984**, *48*, 2901-2903.

(45) Yoshihama, M.; Higashiro, K.; Rao, E. A.; Akedo, M.; Shanabruch, W. G.; Follettie, M. T.; Walker, G. C.; Sinskey, A. J. *J Bacteriol* **1985**, *162*, 591-597.

(46)   Smith, M. D.; Flickinger, J. L.; Lineberger, D. W.; Schmidt, B. *Appl Environ Microbiol* 1986, *51*, 634-639.

(47)   Fernandez-Gonzalez, C.; Cadenas, R. F.; Noirot-Gros, M. F.; Martin, J. F.; Gil, J. A. *J Bacteriol* 1994, *176*, 3154-3161.

(48)   Goyal, D.; Wachi, M.; Kijima, N.; Kobayashi, M.; Yukawa, H.; Nagai, K. *Plasmid* 1996, *36*, 62-66.

(49)   Ehrlich, S. D. In *Mobile DNA*; Berg, D. E., Howe, M. M., Eds.; American Society for Microbiology Press: Washington D.C., 1989; pp 799-832.

(50)   Jannière, L.; Gruss, A.; Ehrlich, S. D. In *Bacillus subtilis and other gram-positive bacteria: biochemistry, physiology, and molecular genetics*; Sonenshein, A. L., Hoch, J. A., Losick, R., Eds.; American Society for Microbiology: Washington D.C., 1993; pp 625-644.

(51)   Blondelet-Rouault, M. H.; Weiser, J.; Lebrihi, A.; Branny, P.; Pernodet, J. L. *Gene* 1997, *190*, 315-317.

(52)   Inui, M.; Nakata, K.; Roh, J. H.; Zahn, K.; Yukawa, H. *J Bacteriol* 1999, *181*, 2689-2696.

(53)   Steinmetz, M.; Le Coq, D.; Djemia, H. B.; Gay, P. *Mol Gen Genet* 1983, *191*, 138-144.

(54)   Gay, P.; Le Coq, D.; Steinmetz, M.; Berkelman, T.; Kado, C. I. *J Bacteriol* 1985, *164*, 918-921.

(55)   Tauch, A.; Gotker, S.; Puhler, A.; Kalinowski, J.; Thierbach, G. *Plasmid* 2002, *48*, 117-129.

(56)   Zhang, Y.; Praszkier, J.; Hodgson, A.; Pittard, A. J. *J Bacteriol* 1994, *176*, 5718-5728.

(57)   Nesvera, J.; Patek, M.; Hochmannova, J.; Abrhamova, Z.; Becvarova, V.; Jelinkova, M.; Vohradsky, J. *J Bacteriol* 1997, *179*, 1525-1532.

(58)   Archer, J. A.; Sinskey, A. J. *J Gen Microbiol* 1993, *139*, 1753-1759.

(59)   Khan, S. A. *Microbiol Mol Biol Rev* 1997, *61*, 442-455.

(60)   del Solar, G.; Giraldo, R.; Ruiz-Echevarria, M. J.; Espinosa, M.; Diaz-Orejas, R. *Microbiol Mol Biol Rev* 1998, *62*, 434-464.

(61)   Ilyina, T. V.; Koonin, E. V. *Nucleic Acids Res* 1992, *20*, 3279-3285.

(62)   Kramer, M. G.; Khan, S. A.; Espinosa, M. *Embo J* 1997, *16*, 5784-5795.

(63)   Yamagata, H.; Terasawa, M.; Yukawa, H. *Catalysis Today* 1994, *22*, 621-627.

(64)   Terasawa, M.; Inui, M.; Goto, M.; Kurusu, Y.; Yukawa, H. *Appl Microbiol Biotechnol* 1991, *35*, 348-351.

(65)   Terasawa, M.; Inui, M.; Goto, M.; Shikata, K.; Imanari, M.; Yukawa, H. *J Ind Microbiol* 1990, *5*, 289-294.

Chapter 12

Cellular Engineering of *Escherichia coli* for the Enhanced Protein Production by Fermentation

S. Y. Lee and K. J. Jeong

Metabolic and Biomolecular Engineering, National Research Laboratory, Department of Chemical and Biomolecular Engineering and BioProcess Engineering Research Center, Korea, Advanced Institute of Science and Technology, 373–1 Guseong-dong, Yuseong-gu, Daejeon 305–701, Korea
*Corresponding Author: e-mail: leesy@mail.kaist.ac.kr

In 1982, approval of FDA for Eli Lilly's recombinant human insulin was just the prelude to the use of *Escherichia coli* as a host strain for recombinant protein production. After two decades, although many alternative organisms such as mammalian and plant cells are now being used for recombinant protein production, *E. coli* still remains the most valuable host for the production of recombinant proteins because of the advantages such as fast growth, well-characterized genetics, availability of numerous vector systems and high cell density culture (HCDC) technologies. In recombinant *E. coli*, once an optimal expression system is constructed, protein production can be enhanced by cellular engineering. Especially, during the HCDC, suitable cellular engineering strategies are required for the increase of the specific productivity of proteins. Here, we review the recent progress in the cellular engineering useful for the enhanced production of recombinant proteins in HCDC of *E. coli*.

Introduction

The primary goal of HCDC research is the cost-effective production of target proteins using high productivity techniques. The volumetric productivity of a recombinant protein will depend on the biomass concentration as well as the specific protein productivity. Whereas fed-batch processes primarily focus on increasing the biomass concentration, the cultivation condition itself can affect the protein productivity. In order to optimize the volumetric recombinant protein productivity, one must consider both of these factors in HCDC. Various HCDC techniques have been developed to increase the productivity of recombinant proteins, and also to provide advantages such as reduced culture volume, enhanced downstream processing, reduced wastewater, lower production costs and reduced investment in equipment (*1*). A number of different strategies were developed to increase the volumetric productivity by the optimization of fermentation conditions including culture temperature (*2*), DO concentration (*3, 4*), medium components (*5*), induction point (*6*), and feeding strategies (*7, 8*). However, these strategies exert in general rather limited influence on the quality of the product. In contrast, cellular engineering strategies such as co-production of proteins to facilitate protein folding and/or secretion, stabilization of gene expression system, and development of *E. coli* mutant to prevent proteolysis and production of byproduct, offer more direct advantages to increase the specific productivity of recombinant proteins. This paper reviews recent advances in cellular engineering of *E. coli* for enhanced protein production.

Control of transcription and translation

For the efficient production of recombinant proteins in *E. coli*, the primary consideration is the choice of suitable expression vector. In general, the expression vector should contain several necessary elements such as origin of replication, antibiotic resistance gene or other selectable marker, promoter and transcription terminators *etc.*

Promoter

A number of promoters are available for the high-level expression of recombinant proteins in *E. coli* and are reviewed extensively elsewhere (*9, 10*).

For many years, the *lac* operon based promoter (*trc, lac, tac, lacUV5-T7 hybrid,* etc.) have widely been used. The pET series of vectors have also been extensively used because they contain a very strong T7 promoter and allow tight regulation of gene expression preventing the basal-level gene expression. However, because of its high cost and toxicity to human, the use of IPTG as a inducer may not be desirable for the large scale production of recombinant proteins, especially therapeutic proteins. Among alternative inducible promoters suitable for large scale production by high cell density culture, arabinose promoter (P_{BAD}), *trp* promoter, temperature inducible promoter (P_L and P_R) and oxygen-dependent promoter (P*nar*), are commonly used. The above promoters along with recently identified promoters and their uses in the protein production are summarized in Table 1.

Table 1. Promotors used for the high-level production of recombinant proteins in *E. coli*

Promoter	Inducer	Protein production			Ref.
		Protein	Max cell density	Protein yield	
araBAD	arabinose	Interferon-α	OD 160	30%[a]	*11*
		Cyclohexanone monooxygenase	5.5 g/L	3500 U/L	*12*
Nar	O₂ limitation	β-galactosidase	OD 176	45%[a]	*13*
glnA	ammonium	CAT protein	OD 100	55 mg/L	*14*
rhaBAD	rhamnose	Carbamoylase	90 g/L	3.8 g/L	*15*
malK	carbohydrate	β-galactosidase	50 g/L	20,000 U/L	*16*
cspA	low temp.	β-galactosidase	OD 40	15%[a]	*17*
recA	nalidixic acid	*B. thuringiensis tenebrionis*	OD 40	85 mg/L	*18*

[a]Content of target protein in total protein

Besides the inducible promoters described above, several constitutive promoters such as the phage T5 promoter, the D-amino acid aminotransferase (D-AAT) promoter (HCE) of *Geobacillus toebii*, and the *Staphylococcus aureus* protein A promoter (P_{SPA}) have also been used for recombinant protein production by HCDC (*19, 20, 21*). Poo et al (*20*) used a new constitutive promoter (HCE) of *Geobacillus toebii* for the production of human tumor necrosis factor-α (rhTNFα). During the HCDC, the HCE system allowed the higher expression level (4 g rhTNFα /L) and higher cell density (20 g DCW/L) than those obtained with the pET series vector (3.9 g/L and 10 g DCW/L, respectively). Even though there are some drawbacks such as decrease of cell growth rate due to the metabolic burden during the early stage of cell growth, the use of constitutive promoter still provides several advantages. Since no inducer needs to be added, one can save the cost of inducer and the time to optimize the induction condition.

mRNA stability

mRNA stability is one of the important factors to achieve high-level gene expression by affecting translation. The decay rate of mRNA transcript is mediated by ribonuclease (PNPase, RNase E, II, III, and etc.). Several other factors including the amount of ribosome, secondary structure of mRNA and translation rate can also affect mRNA stability (22). Smolke et al (23, 24) examined the effect of introducing mRNA stability control elements (5' or 3' secondary structures, RNase cleavage sites, and gene location) into various regions of mRNA. The protein production level could be changed from 1- to 1000-fold depending on the combination of control elements.

In *E. coli*, bacteriophage T7 promoter has widely been used for the overexpression of target gene because T7 RNA polymerase can transcribe very fast, which results in the high-level production of target protein. However, this fast transcription also leads to asynchronous transcription and translation, which causes reduction in protein expression. Iost and Dreyfus (25) showed that the co-expression of *E. coli* DEAD-box protein enhanced production of β-galactosidase (30-fold) when using the T7 promoter. The helicase activity of DEAD-box system to bind and dissociate RNA duplexes in an ATP-dependent manner makes it a good candidate as an RNA-protecting molecule for unstable transcripts produced by the T7 expression system. When the DEAD-box protein was employed in the production of the tandem multimer (36 repeats) of antimicrobial peptide, buforin II, much higher level of expression could be achieved using the T7 promoter compared with less strong *tac* promoter (26).

Codon usage

The codon usages in the genes vary among the different organisms. It has been shown that human genes that contain codons which are rare in *E. coli* may be inefficiently expressed (27). A subset of codons, AGG/AGA (Arg, 0.14%), CGA (Arg, 0.31%), AUA (Ile, 0.41%), CUA (Leu, 0.32%), and CCC (Pro, 0.43%) are the least used codons in *E. coli*, which agrees with the finding that the tRNAs for these codons are rare in *E. coli* (27). The presence of rare codons in the target gene results in low-level translation of the target gene or translation error such as mistranslation and premature termination by frameshift, and consequently leads to low productivity of target protein or sometimes processive degradation. To minimize the effect of codon usage difference, two alternative strategies have been used. The first is the alteration of rare codons based on the preferred codon usage of *E. coli* (*28*), and the second is the co-expression of genes that encode the rare tRNAs such as *argU* (*dnaY*) gene that encodes the tRNA^{Arg} (*29*). Recently, *E. coli* strain having extra copies of *argU*, *ileY* and *leuW* tRNA genes in chromosome was developed. When this strain was used for the expression of a gene containing rare codons, much higher level of recombinant protein could be achieved compared with wild type *E. coli* host (*30, 31*). Some mRNAs having high GC contents at the 5' coding region can affect its translation, which can also be overcome by the alteration of codon usage (*32, 33*). The surprisingly great impact of this strategy has been demonstrated by Libessart et al (*34*) who reported that only a single codon change could increase the expression level of branching enzyme II by 7-fold compared with the native enzyme.

Ribosome pool

Recombinant protein production has been shown to elicit various stress responses in *E. coli*. One of those problems is inefficient translation caused by the decrease of ribosome pool. Generally, it is suggested that guanosine tetraphosphate (ppGpp) affects RNA polymerase selectivity, rendering it unable to initiate transcription at stable RNA promoters. Also ppGpp itself is a potent inhibitor of protein synthesis by reducing the RNA chain growth rate (*35*). The level of ppGpp can also be used as a signal for examining the metabolic load on host cell resulting from the recombinant protein production (*36*). Dedhia et al (*37*) showed that elimination of intracellular ppGpp by using the ppGpp-deficient *E. coli* mutant (*spoT*, *relA*) has resulted in a five-fold increase in the levels of five recombinant proteins during fed-batch fermentations. FIS, a heat stable-DNA binding protein (11.2 kDa) in *E. coli*, is also known to stimulate stable RNA synthesis under conditions of nutrient upshift. Its overexpression during exponential growth phase activated *rrn* promoters and consequently led to improved production of recombinant protein during fermentation (*38*).

Plasmid copy number

Traditionally, high-copy number (several hundreds copies per cell) plasmids such as pUC vectors have been used for the overproduction of recombinant proteins. However, this high-copy number may cause the instability of plasmid and decrease of cell growth rate, and may also prevent the soluble protein production and secretion of premature proteins. Jones and Keasling (39) developed the low-copy plasmid based on F plasmid of E. coli which are maintained at 1-2 copies per cell. This plasmid was found to be stable in the absence of selection pressure and to confer a low metabolic burden on the host. Also, integration of expression cassette into chromosome is considered as an alternative method to decrease the gene copy number (40). This strategy can prevent segregational instability of plasmid and use of strong promoter can also enhance the protein production without aggregation as inclusion bodies.

In addition to promoter, mRNA stability, ribosome pool, codon usage and plasmid copy number, there are many more factors to be considered for the optimal recombinant production. Some of the important ones include plasmid stability (*41, 42*), and multimerization (tandem repeats) of a target gene (*26, 43*), which have been reviewed elsewhere (*9, 10, 44*).

Protein targeting into three different compartments of *E. coli*

E. coli, as it being a gram-negative bacterium, has two membranes which make 3 different compartments (cytoplasm, periplasm and culture medium). The decision to target a protein to a specific cellular compartment rests on balancing the advantages and disadvantages of expressing the protein at each compartment (*9, 45*).

Production in cytoplasm

Overproduction of recombinant proteins in the cytoplasm is frequently accompanied by their misfolding and segregation into insoluble aggregates known as inclusion bodies. Formation of inclusion bodies provides several advantages such as easy isolation of protein produced with high purity by simple centrifugation, protection from proteolysis, relief of potential toxicity of functional recombinant protein, and high level production yield by HCDC (6). However, refolding of recombinant protein often results in the remarkable reduction in the yield, and furthermore, some proteins can not be refolded. Therefore, if the recombinant protein is resistant to proteolysis by cytoplasmic proteases and is difficult to refold after denaturation, it will be desirable to produce the protein as soluble form in the cytoplasm. There are several different strategies to minimize inclusion body formation as follows: (i) reduction of the protein synthesis rate by controlling the promoter strength, culture temperature and amount of inducer (*9, 17*), (ii) use of fusion partner such as thioredoxin,

DsbA mutant form, NusA and bacteriophage head protein D (gpHD) (*46, 47, 48*), (iii) co-expression of molecular chaperone and/or foldase including GroELS, DnaK, DnaJ and trigger factor (*49, 50, 51*), and (iv) use of *E. coli* mutant in which less reducing environment is provided (*26, 52, 53, 54*). Decrease of culture temperature is a traditional method for the enhancement of soluble protein production and, in this sense, use of cold-responsive promoter (λP_L and cspA promoter) is considered as one of the useful method for the soluble protein production. Vasina et al (17) examined the usefulness of fermentation and strain engineering approaches in circumventing the problem of cspA promoter. An rbfA null mutant strain allowed constitutive protein expression for up to 7 hours after temperature downshift (23°C) and their performance was validated in high density fed-batch fermentations. Various fusion partner have been widely used because of combined advantage of soluble protein production as well as high-level production and affinity purification (9, 10). However, this strategy requires the trial-and-error procedure for the choice of proper fusion partner. Davis et al (47) suggested the new approach to solve this drawback. Among ~4000 proteins of E. coli, they chose a NusA protein as a fusion partner based on the statistical solubility modeling and the its use for the production of human IL-3, bovine growth hormone and human IFN-γ showed much higher solubility of target protein than that by use of other fusion partner such as thioredoxin. In the cytoplasm of E. coli, disulfide bond can be formed by the action of two thioredoxins (TrxA and TrxC) and three glutaredoxins. However, they are usually present in a reduced state by the action of thioredoxin reductase (TrxB) and glutathion (gshA and gshB), which prevent the formation of disulfide bonds. Bessette et al (53) developed E. coli mutant strain (FA113) which grow well in spite of impairment in the reduction of both thioredoxin and glutathione (trxB gor supp). Several model proteins including highly complex protein (full length tPA) containing 17 disulfide bonds were examined using this system. Appreciable yields of active proteins could be obtained by the co-expression of TrxA variant or leaderless DsbC, which exceeded the yields that could be obtained by periplasmic expression. Venturi et al (54) also reported that a higher level production of fully functional Fab antibody could be achieved in the same E. coli mutant. These results indieicate that *E. coli* cytoplasm may now be considered as an alternative space for the preparative production of soluble and folded protein of complex structure. Recently, several heat shock proteins including ClpB, HtpG and IbpA/B have also been reported for their abilities to act as chaperones (*55, 56, 57*), which may be useful for recombinant protein production.

Secretion into the periplasm

The periplasm of *E. coli* is an attractive space for the production of recombinant proteins because (i) it provides an oxidizing environment which can facilitate proper folding of proteins, (ii) it contains less proteases than cytoplasm, and thus reducing protein degradation, and (iii) there are less cellular proteins which is beneficial for the purification of recombinant proteins. Approximately 20% of proteins synthesized in *E. coli* are present outside of cytoplasm. Translocation of recombinant proteins into periplasm, culture medium or inner/outer membrane can be mediated by four different pathways: (i) Sec-dependent general secretory pathway (*58*), (ii) signal recognition particle (SRP) dependent pathway (*59*), (iii) YidC-dependent process (*60*), and (iv) twin-arginine translocation (Tat) pathway (*61*). Among these 4 distinct pathways, the Sec-dependent pathway has been widely used for the secretory production of proteins in *E. coli*. However, more interest is building up in the use of the other pathways. Especially, the Tat pathway discovered from the thylakoid membrane of photosynthetic organisms seems to be a potentially useful system because of its distinctive ability to transport the folded proteins across the cytoplasmic membrane (*62*). Although, only two model proteins (green fluorescent protein and *Thermus thermophilus* alkaline phosphatse) have so far been reported to be successfully translocated (*63, 64, 65*), more examples are expected to appear in the future.

For the secretory production, fusion of signal peptide to the N-terminus of mature target protein is required. Various signal peptides such as OmpA, LamB, OmpF, PelB, PhoA, SpA, ST-II, and β-lactamase have widely been used (*9, 66*). The choice of signal peptide is important for the efficient secretion of recombinant proteins because some signal peptides do not always ensure successful secretion of recombinant proteins due to several reasons including autolytic activities of weakened outer membrane, the low product levels, incomplete processing, and different characteristics of the proteins to be secreted in *E. coli*. The combination of signal peptide and mature protein can affect their secondary/tertiary structure of fusion protein which may facilitate or prevent the translocation of recombinant proteins across the membrane. Therefore, the following strategies are often considered for the efficient translocation and proper processing: (i) co-expression of molecular chaperone such as DnaKJ and GroELS (*67*), (ii) co-expression of proteins participating in membrane-transport process including SecB (*68*), (iii) co-expression of signal peptidase (*69*), (iv) the use of *E. coli* mutant such as *prlF*-deficient strain (*70*), and (v) the use of fusion partner including DsbA and thioredoxin (*71, 72*).

In the secretory protein production in *E. coli*, it is obviously important to produce proteins in corrected folded form. However, in many cases, misfolding and aggregation of secreted proteins in periplasm are observed. One strategies for correct folding of secreted proteins is the supply of components involved in protein folding such as periplasmic foldase. The foldases related to the protein folding in the periplasm of *E. coli* are summarized in Table 2.

Table 2. Foldases used for the soluble protein production in the periplasm of *E. coli*

Foldase	Role
DsbA	Main catalyst for oxidation of protein and peptide secreted into periplasm
DsbC	Isomerization of incorrect disulfide bonds (oxidoreductase)
DsbE	Disulfide oxidoreductase
DsbG	Isomerization of disulfide bond (chaperon-like activity)
PpiA	Prevention of improper interaction
PpiD	Catalysis of prolyl peptide bond isomerization
Skp	Maintenance of early intermediate of outer membrane proteins
Fkp	Prevention of improper interaction
SurA	Catalysis of prolyl peptide bond isomerization

Dsb proteins are the most well-known foldases and have been widely used for the correct folding of secreted proteins. In general, DsbA and DsbC play distinct roles in disulfide bond formation and disulfide bond isomerization, respectively. Co-expression of these foldases with or without its regenerator (DsbB and DsbD, respectively) has led to the successful folding of secreted proteins (*73, 74, 75*). Qiu et al (*73*) showed that the *dsbC* co-expression system could greatly enhance the folding of tPA, which has 17 disulfide bonds, during HCDC (15 L fermentor). Bothmann and Pluckthun (*76, 77*) developed a search system for periplasmic factors improving phase display, and by using this system, two factors, Skp containing the chaperone like activity, and Fkp, one of the peptidylprolyl *cis,trans*-isomerase (PPIase), were identified. Co-expression of these foldases improved the folding of a number of aggregation-prone single chain antibody fragment. In addition, there are several other PPIase including SurA, PpiA and PpiD which can also assist the folding of secreted protein (*52, 78*). Zhan et al (*79*) reported interesting results that the co-expression of yeast disulfide isomerase also facilitated the disulfide bonds formation of tPA and bovine pancreatic trypsin inhibitor (BPTI) in the *E. coli* periplasm.

Excretion into culture medium

The excretory production of recombinant proteins into the culture medium has several significant advantages: the least level of proteolysis, simple purification, improved protein folding and N-terminal authenticity. Due to these advantages, there have been several attempts to develop strategies for excretory production of proteins as follows: (i) use of fusion partners such as *pelB* leader, the *ompA* leader, the protein A leader, alkaline phosphatase (PhoA) leader and maltose binding protein (*80, 81*), (ii) use of dedicated translocators such as hemolysin and pullulanase system (*82, 83*), (iii) co-expression of *kil* gene, *tolAIII* gene, bacteriocin release protein (BRP) gene and the mitomycin-induced bacteriocin release protein gene (*84*), and (iv) use of leaky L-form *E. coli* cells that release periplasmic proteins into the culture medium due to the loss of outer membrane integrity (*85*). However, these strategies showed the relatively high contamination of cellular proteins in culture supernatant because of destabilization of the outer membrane. Especially, L-form *E. coli* cells are hypersensitive to detergents and EDTA, and cannot be cultivated to a high density typically required for the efficient production of recombinant proteins (*86*). In spite of superior advantages, its use for HCDC have many problems to be solved yet. Recently, Jeong and Lee (*87*) found that *E. coli* BL21 strains were able to excrete a large amount of OmpF into culture medium during HCDC. From this interesting phenomenon, a new efficient excretion system using OmpF protein as a fusion partner was developed which is suitable for the large scale excretory production of recombinant proteins. By using this system, up to 15 g of OmpF-β-endorphin fusion protein was excreted into culture medium with a high purity (75%).

Conclusions

As reviewed above, various cellular engineering strategies have been successfully employed for the enhanced production of recombinant by *E. coli*. During the past several years, there have been great advances in genomics and proteomics tools and, by using these tools, global analysis of cellular responses caused by high-level production of recombinant proteins during HCDC has been reported (*88, 89, 90, 91*). The results obtained from the proteome and transcriptome analysis may give the insights into the global regulation networks and interaction of cellular proteins during the high-level protein production in HCDC, and consequently, can suggest new cellular and metabolic engineering strategies for much more efficient production of recombinant proteins. Even though several problems remain in the production of complex proteins and proteins requiring post-translational modifications, these problems are expressed to be solved by new cellular engineering strategies. All these effort will strengthen the dominant status of *E. coli* as a host strain for recombinant protein production.

Acknowledgements

Our work described in this paper was supported by the Basic Industrial Research Program of the Ministry of Commerce, Industry and Energy, and by the National Research Laboratory program of the Ministry of Science and Technology. Further support from Center for Ultramicrochemical Process Systems is also appreciated. KJJ is postdoctoral fellow supported by Brain Korea 21 project.

References

1. Lee, S.Y. *Trends Biotechnol.* **1996**, 14, 98-105.
2. Gupta, P.; Sahai, V.; Bhatnagar, R. *Biochem. Biophys. Res. Commun.* **2001**, 285, 1025-1033.
3. Hewitt, C. J.; Nebe-Von Caron, G.; Axelsson, B.; McFarlane, C. M.; Nienow, A. W. *Biotechnol. Bioeng.* **2000**. 70, 381-390.
4. Johnston, W.; Cord-Ruwisch, R.; Cooney, M. J. *Bioprocess Biosyst. Eng.* **2002**, 25, 111-120.
5. Riesenberg, D.; Schulz, V.; Knorre, W. A.; Pohl, H. D.; Korz, D.; Sanders, E. A.; Ross, A.; Deckwer, W. D. *J. Biotechnol.* **1991**, 20, 17-27.
6. Jeong, K. J.; Lee, S. Y. *Appl. Environ. Microbiol.* **1999**, 65, 3027-3032.
7. Wong, H. H.; Kim, Y. C.; Lee, S. Y.; Chang, H.N. *Biotechnol. Bioeng.* **1998**, 60, 271-276.
8. Wang, F. S.; Cheng, W. M. Biotechnol. Prog. **1999**, 15, 949-952.
9. Markrides, S. C. *Microbiol. Rev.* **1996**, 60, 512-538.
10. Balbas, P. *Mol Biotechnol.* **2001**, 19, 251-267.
11. Lim, H.K.; Jung, K.H.; Park, D.H.; Chung, S.I. *Appl. Microbiol. Biotechnol.* **2000**, 53, 201-208.
12. Doig, S. D.; O'Sullivan, L. M.; Patel, S.; Ward, J. M.; Woodley, J. M. *Enzyme Microb. Technol.* **2001**, 28, 265-274.
13. Han, S. J.; Chang, H. N.; Lee, J. *Biotechnol. Bioeng.* **2001**, 72, 573-576.
14. Schroeckh V, Wenderoth R, Kujau M, Knupfer U, Riesenberg D. *J. Biotechnol.* **1999**, 75, 241-250.

15. Wilms, B.; Hauck, A.; Reuss, M.; Syldatk, C.; Mattes, R.; Siemann, M.; Altenbuchner, J. *Biotechnol. Bioeng.* **2001**, 73, 95-103.
16. Larsson, G.; Bostrom, M. *Appl. Microbiol. Biotechnol.* **2002**, 59, 231-238.
17. Vasina, J. A.; Peterson, M. S.; Baneyx, F. *Biotechnol. Prog.* **1998**, 14, 714-721.
18. Gustafson, M. E.; Clayton, R. A.; Lavrik, P. B.; Johnson, G. V.; Leimgruber, R. M.; Sims, S. R.; Bartnicki, D. E. *Appl. Microbiol. Biotechnol.* **1997**, 47, 255-261.
19. Park. S. J.; Georgiou, G.; Lee, S. Y. *Biotechnol. Prog.* **1999**, 15, 164-167.
20. Poo, H.; Song, J. J.; Hong, S. P.; Choi, Y. H.; Yun, S. W.; Kim, J. H.; Lee, S. C.; Lee, S. G.; Sung, M. H. *Biotechnol. Lett.* **2002**, 24, 1185-1189.
21. Chauhan, V.; Singh, A.; Waheed, S. M.; Singh, S.; Bhatnagar, R. *Biochem. Biophys. Res. Commun.* **2001**, 283, 308-315.
22. Carrier, T. A.; Keasling, J. D. *Biotechnol. Prog.* **1997**, 13, 699-708.
23. Smolke, C. D.; Carrier, T. A.; Keasling, J. D. *Appl. Environ. Microbiol.* **2000**, 66, 5399-5405.
24. Smolke, C. D.; Keasling, J. D. *Biotechnol. Bioeng.* **2002**, 78, 412-424.
25. Iost, I; Dreyfus, M. *Nature.* **1994**, 372, 193-196.
26. Lee, J. H.; Kim, M. S.; Cho, J. H.; Kim, S. C. *Appl. Microbiol. Biotechnol.* **2002**, 58, :790-796.
27. Kane, J. F. *Curr. Opin. Biotechnol.* **1995**, 6, 494-500.
28. Hu, X.; Shi, Q.; Yang, T.; Jackowski, G. *Protein Expr. Purif.* **1996**, 7, 289-293.
29. Dieci, G.; Bottarelli, L.; Ballabeni, A.; Ottonello, S. *Protein Expr. Purif.* **2000**, 18, 346-354.
30. Kleber-Janke, T.; Becker, W. M. *Protein Expr. Purif.* **2000**, 19, 419-424.
31. Laine, S.; Salhi, S.; Rossignol, J. M. *J. Virol. Methods.* **2002**, 103, 67-74.
32. Pedersen-Lane, J.; Maley, G. F.; Chu, E.; Maley, F. *Protein Expr. Purif.* **1997**. 10, 256-262.
33. Jeong, K. J.; Lee, S. Y. *Protein Expr. Purif.* **2001**, 23, 311-318.
34. Libessart, N.; Preiss, J. *Protein Expr. Purif.* **1998**, 14, 1-7.
35. Svitil, A.L.; Cashel, M.; Zyskind, J.W. *J. Biol. Chem.* **1993**, 268, 2307-2311.
36. Cserjan-Puschmann, M.; Kramer, W.; Duerrschmid, E.; Striedner, G.; Bayer, K. *Appl. Microbiol. Biotechnol.* **1999**, 53, 43-50.
37. Dedhia, N.; Richins, R.; Mesina, A.; Chen, W. *Biotechnol. Bioeng.* **1997**, 53, 379-386.
38. Richins, R.; Hyay, T.; Kallio, P.; Chen, W. *Biotechnol. Bioeng.* **1997**, 56, 138-144.
39. Jones, K. L.; Keasling, J. D. *Biotechnol. Bioeng.* **1998**, 59, 659-665.
40. Marchand, I.; Nicholson, A. W.; Dreyfus, M. *Gene.* **2001**, 262, 231-238.
41. Ansorge, M. B.; Kula, M. R. *Appl. Microbiol. Biotechnol.* **2000**, 53, 668-673.

42. Horn, U.; Strittmatter, W.; Krebber, A.; Knupfer, U.; Kujau, M.; Wenderoth, R.; Muller, K.; Matzku, S.; Pluckthun, A.; Riesenberg, D. *Appl. Microbiol. Biotechnol.* **1996**, 46, 524-532.

43. Kim, Y. C.; Kwon, S.; **Lee, S. Y.;** Chang, H. N. *Biotechnol. Lett.* **1998**, 20, 799-803.

44. Swartz, J. R. *Curr. Opin. Biotechnol.* **2001**, 12, 195-201.

45. Jonasson, P.; Liljeqvist, S.; Nygren, P. A.; Stahl, S. *Biotechnol. Appl. Biochem.* **2002**, 35, 91-105.

46. Zhang, Y.; Olsen, D. R.; Nguyen, K. B.; Olson, P. S.; Rhodes, E. T.; Mascarenhas, D. *Protein Expr. Purif.* **1998**, 12,159-165.

47. Davis, G. D.; Elisee, C.; Newham, D. M.; Harrison, R. G. *Biotechnol. Bioeng.* **1999**, 65, 382-388.

48. Forrer, P.; Jaussi, R. *Gene.* **1998**, 224, 45-52.

49. Widersten, M. *Protein Expr. Purif.* **1998**, 13, 389-395.

50. Lamark, T.; Ingebrigtsen, M.; Bjornstad, C.; Melkko, T.; Mollnes, T. E.; Nielsen, E. W. *Protein Expr. Purif.* **2001**, 22, 349-358.

51. Nishihara, K.; Kanemori, M.; Yanagi, H.; Yura, T. *Appl. Environ. Microbiol.* **2000**, 66, 884-889.

52. Levy, R.; Weiss, R.; Chen, G.; Iverson, B. L.; Georgiou, G. *Protein Expr. Purif.* **2001**, 23, 338-347.

53. Bessette, P. H.; Aslund, F.; Beckwith, J.; Georgiou, G. *Proc. Natl. Acad. Sci. USA.* **1999**, 96, 13703-13708.

54. Venturi, M.; Seifert, C.; Hunte, C. *J. Mol. Biol.* **2002**, 315, 1-8

55. Kitagawa, M.; Miyakawa, M.; Matsumura, Y.; Tsuchido, T. *Eur. J. Biochem.* **2002**, 269, 2907-2917.

56. Thomas, J. G.; Baneyx, F. *Mol. Microbiol.* **2000**, 36, 1360-1370.

57. Hoffmann, F.; Rinas, U. *Biotechnol. Prog.* **2000**, 16, 1000-1007.

58. Pugsley. A. P. *Microbiol. Rev.* **1993**, 57, 50-108.

59. Meyer, D. I.; Krause, E.; Dobberstein, B. *Nature* **1982**, 297, 647-650.

60. Samuelson, J. C.; Chen, M.; Jiang, F.; Moller, I.; Wiedmann, M.; Kuhn, A.; Phillips, G. J.; Dalbey, R. E. *Nature.* **2000**, 406, 637-641.

61. Berks, B. C. *Mol. Microbiol.* **1996**, 22, 393-404.

62. Berks, B. C.; Sargent, F.; Palmer, T. *Mol. Microbiol.* **2000**, 35, 260-274.

63. Angelini, S.; Moreno, R.; Gouffi, K.; Santini, C.; Yamagishi, A.; Berenguer, J.; Wu, L. *FEBS Lett.* **2001**, 506, 103-107.

64. Thomas, J. D.; Daniel, R. A.; Errington, J.; Robinson, C. *Mol. Microbiol.* **2001**, 39,:47-53.

65. Santini, C. L.; Bernadac, A.; Zhang, M.; Chanal, A.; Ize, B.; Blanco, C.; Wu, L. F. *J. Biol. Chem.* **2001**, 276, 8159-8164.

66. Holland, B.; Chervaux, C. *Curr. Opin. Biotechnol.* **1994**, 5, 468-474.

67. Schaffner, J.; Winter, J.; Rudolph, R.; Schwarz, E. *Appl. Environ. Microbiol.* **2001**, 67, 3994-4000

206

68. Chou, C. P.; Tseng, J. H.; Kuo, B. Y.; Lai, K. M.; Lin, M. I.; Lin, H. K. *Biotechnol. Prog.* **1999**, 15, 439-445.
69. Smith, A. M.; Yan, H.; Groves, N.; Dalla Pozza, T.; Walker, M. J. *FEMS Microbiol. Lett.* **2000**, 191, 177-182.
70. Snyder, W. B.; Silhavy, T. J. *J. Bacteriol.* **1992**, 174, 5661-5668.
71. Winter, J.; Neubauer, P.; Glockshuber, R.; Rudolph, R. *J. Biotechnol.* **2001**, 84, 175-185.
72. Wang, Y.; Jing, L.; Xu, K. *J. Biotechnol.* **2002**, 94, 235-244.
73. Qiu, J.; Swartz, J. R.; Georgiou, G. *Appl. Environ. Microbiol.* **1998**, 64, 4891-4896.
74. Jeong, K. J.; Lee, S. Y. *Biotechnol. Bioeng.* **2000**, 67, 398-407.
75. Kurokawa, Y.; Yanagi, H.; Yura, T. *J. Biol. Chem.* **2001**, 276, 14393-14399.
76. Bothmann, H.; Pluckthun, A. *Nat. Biotechnol.* **1998**, 16, 376-380.
77. Bothmann, H.; Pluckthun, A. *J. Biol. Chem.* **2000**, 275, 17100-17105
78. Mavrangelos, C.; Thiel, M.; Adamson, P. J.; Millard, D. J.; Nobbs, S.; Zola, H.; Nicholson, I. C. *Protein. Expr. Purif.* **2001**, 23, 289-295.
79. Zhan, X.; Schwaller, M.; Gilbert, H. F.; Georgiou, G. *Biotechnol. Prog.* **1999**, 15, 1033-1038.
80. Xu, R.; Du, P.; Fan, J. J.; Zhang, Q.; Li, T. P.; Gan, R. B. *Protein Expr. Purif.* **2002**, 24, 453-459.
81. Ko, J. H.; Park, D. K.; Kim, I. C.;. Lee, S. H.; Byun, S. M. *Biotechnol. lett.* **1995**, 17, 1019-1024.
82. Blight, M. A.; Holland, I. B. *Trends Biotechnol.* **1994**. 12, 450-455.
83. Kern, I.; Ceglowski, P. Gene. **1995**, 163, 53-57.
84. Wan, E. W.; Baneyx, F. *Protein Expr. Purif.* **1998**, 14, 13-22.
85. Gumpert, J.; Hoischen, C. *Curr. Opin. Biotechnol.* **1998**, 9, 506-509.
86. Onoda, T.; Enokizono, J.; Kaya, H.; Oshima, A.; Freestone, P.; Norris, V. *J. Bacteriol.* **2000**, 182, 1419-1422.
87. Jeong, K. J.; Lee, S. Y. *Appl. Environ. Microbiol.* **2002**, 68, 4979-4985.
88. Franzen, B.; Becker, S.; Mikkola, R.; Tidblad, K.; Tjernberg, A.;, Birnbaum, S.; *Electrophoresis.* **1999**, 20, 790-797.
89. Noronha, S. B.; Yeh, H. J.; Spande, T. F.; Shiloach, J. *Biotechnol. Bioeng.* **2000**, 68, 316-327.
90. Gill, R. T.; DeLisa, M. P.; Valdes, J. J.; Bentley, W. E. *Biotechnol. Bioeng.* **2001**, 72, 85-95
91. Oh, M. K.; Liao, J. C. *Metab. Eng.* **2000**, 2, 201-209.

Chapter 13

Metabolic Engineering of Folic Acid Production

T. Zhu[1], R. Koepsel[1], M. M. Domach[2], and M. M. Ataai[1]

[1]Department of Chemical and Petroleum Engineering, University
of Pittsburgh, Pittsburgh, PA 15261
[2]Department of Chemical Engineering, Carnegie Mellon University,
Pittsburgh, PA 15213

Computer-aided metabolic flux analysis has suggested how to
divert cell raw materials to elevate the production of folic acid
in *E. coli*. One of the strategies, pyruvate kinase (PYK)
deletion, is predicted to divert resources to increase folic acid
production. This flux redirecting strategy is also consistent
with elevating the precursors, phosphoenolpyruvate and
erthyrose-4-phosphate, which provides the mass action
potential to increase the product formation rate. Experimental
measurement of folic acid released in the culture medium
shows that deletion of pyruvate kinase activity from *E. coli*
significantly increases folic acid production. These
calculations and experimental results suggest that PYK
deletion may be a good initial starting point for further
enhancing folic acid production via metabolic engineering.

Introduction

Although digestive system is provided with folate compounds by bacteria and dietary sources, the US FDA has recommended that American diets be supplemented with folic acid to reduce the incidence of certain birth defects and for other health-related reasons. The live stock industry also relies on folic acid supplementation of animal feed. The bulk of commercially sold folic acid is produced by chemical synthesis. Despite the high cost of materials in the chemical synthesis methods and the low product yield, folic acid is not produced commercially by using bacterial cultures. The alternative to biotransformation is complete synthesis by microbes. Here, folic acid is produced from the precursors, erthyrose-4-phosphate (E4P), phosphoenol pyruvate (PEP), and glutamate/glutamine (see Fig. 1). The aromatic amino acid pathway is used up to the intermediate, chorismate. The metabolic capacity of bacterial for production of folic acid is high. Additionally, unlike synthetic folic acid, the glutamated *and* reduced form of the compound is produced by microbes, which is the form the digestive system readily absorbs and the need for reduction to the biologically active form is eliminated.

Linear Programming has been implemented effectively to analyze the productive capabilities of metabolic networks (*1, 2, 3*). An objective is posed such as maximization of a particular flux that leads to the synthesis of an economically valuable product. The answer provided informs one of the yield horizons. Moreover, the values of the fluxes in the optimal solution indicate how to engineer the trafficking of metabolites by altering the available reaction paths and feedback loops.

While very useful, the system of equations and constraints is often underdetermined, which means more than one solution may exist that could satisfy the objective. The identification of other solutions is of interest for several reasons. First, the different solutions may differ in how easy it is to implement them. For example, one solution may indicate more genetic manipulation is required than another that could yield the same value of the objective. Second, there is fundamental value in knowing how many alternatives exist. This provides some insights on how redundant and robust a metabolic system can be. Finally, knowing the portfolio of alternatives ahead of time can result in the development of a more inclusive and tighter patent strategy.

Recent work in metabolic engineering has drawn from the network analysis field to address the problem of enumerating alternate metabolite trafficking solutions. Lee et al (*4*) developed a mixed integer linear programming (MILP) approach to find all the alternate optima in LP models. To extend the prior MILP work, we have developed a software tool (*5*), *MetaboLogic*, to exploit features that enable the linkage of different computational engines by a graphical user interface (GUI). This software automatically generates the mathematical

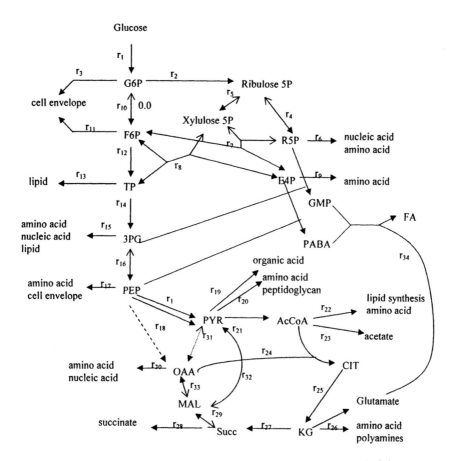

Figure 1. *The metabolic network of E. coli. The two-way arrows represent reversible fluxes with the bold heads showing the constrained net directions. Abbreviation: AcCoA, acetyl-CoA; CIT, citrate; E4P, erythrose-4-phosphate; F6P, fructose-6-phosphate; FA, folic acid; G6P, glucose-6-phosphate; GMP, guanosine monophosphate; KG, α-ketoglutarate; MAL, malate; OAA, oxaloacetate; PEP, phosphoenolpyruvate; 3PG, 3-phosphoglycerate; PABA, para aminobenzoic acid; PYR, pyruvate; R5P, ribose-5-phosphate; Ru5P, ribulose-5-phosphate; Succ, succinyl-CoA; TP, triose phosphate; Xu5P, xylulose-5-phosphate.*

formulation based on the network "picture" the user has *drawn* on a computer screen. All metabolic trafficking scenarios that optimized the objective function are generated.

In this chapter, *MetaboLogic* is used to explore metabolite trafficking options for enhanced folic acid production. Then, the flux distributions of several potentially high yielding metabolic mutants are compared to deduce a potentially viable genetic engineering strategy for enhancing folic acid production. The second part of this chapter will focus on experimental verification of some of the model predictions. We will show that in accordance to the model prediction, pyruvate kinase (PYK) mutation is a promising metabolic engineering starting point for enhancing folic acid production.

Stoichiometric Model and Methods

Model and Analysis Methods

The biochemical reactions in central carbon metabolism of *E. coli* as well as candidate reversible reactions are shown in Figure 1. The network is based on our prior work (*4*).

The metabolite balance equations and constraints have been previously described (*4*). Cell mass compositions data for *E. coli* are from previous work (*6, 7*). Other constraints are derived from NADPH and minimum ATP requirements (*8*). A reference specific growth rate equal to 0.4 h^{-1} is used to generate flux units that can be scaled when other growth rates are considered.

The overall stoichiometry for folic acid production can be described as

$$2PEP + E4P + 1.5\ 3GP + KG + 11\ ATP + R5P + 4\ NADPH \rightarrow$$
$$FA + Pyr + GLdh + 1.5CO2$$

where PEP (phosphoenolpyruvate), E4P (erythrose-4-phosphate), 3GP (3-phosphoglyceric acid), KG (oxoglutarate), R5P (ribose-5-phosphate), Pyr (pyruvate) and Gycolaldehyde (GLdh) are the intermediate metabolites present in the central carbon metabolic pathway. FA (folic acid) and CO_2 are cellular products. This lumped reaction is added to the model, assuming the leakage of intracellular metabolites in this pathway is negligible when the strain is optimized for folic acid production.

Using the *MetaboLogic* model construction tool, all the reactants, reactions and constraints are inputted in less than an hour. The objective function, maximization of folic acid production, is used to find all the flux scenarios that lead to high folic acid production. Computation takes less than one minute on a PIII-900 Hz computer.

Experimental Methods

Cells and Growth Medium

The *E. coli* wild-type strain (JM101) and mutant strain (PB25) lacking activities of both PYKI and PYKII was used in these experiments. The strain was generously provided to us by Dr. Fernando Vale (9). The medium was M9 (10). An initial glucose concentration of 4 g/L was used. Optical density was measured off-line using a Lambda 6 Perkin-Elmer spectrophotometer (Perkin-Elmer, Norwalk, CT), (1 OD660 = 0.36 g cell dry weight/l).

Cultivation & Measurement of Folic Acid

250 mL Shake flasks were used in the cultivation experiments. Folic acid concentration is measured using microbial method (Difco Manual 11th). Samples were collected at different culture time for *E. coli* wild type and mutant culture. These samples are filtered thought 0.2 μm filter, and diluted 40 to 200 times based the estimated folic acid concentration. Stock cultures of the test organism, *L. casei subsp. Rhamnosus* ATCC® 7469 were prepared by stabbing inoculation into prepared tubes of Lactobacilli Agar AOAC. The cultures were incubated at 35-37°C for 18-24 hours. The cultures were stored in the refrigerator at 2-8°C. Transfers were made monthly. The inoculums for assay were prepared by subculturing from a stock culture of *L. casei subsp. rhamnosus* into a tube containing 10 ml Micro Inoculum Broth. These inoculums were incubated at 35-37°C for 16-18 hours. Under aseptic conditions, the tubes were centrifuged to sediment the cells and the supernatant is decanted. Cells were washed 3 times in sterile single-strength Folic Acid Casei Medium. After the third washing, the cells were resuspended in 10 ml sterile single-strength medium and were diluted 100 times. 20 μL of this suspension was used to inoculate each of the assay tubes. In each test tube, 2.5 mL double-strength Folic Acid Casei Medium, 0.5 mL of sample and 2 mL distilled water were added to

the test tube to make the total volume 5 mL. The growth response of the assay tubes was read turbidimetrically after 18-24 hours incubation at 35-37°C. It is essential that a standard curve be constructed for each separate assay. Autoclave and incubation conditions can influence the "standard" curve obtained and absolute values of folic acid concentration can not always be duplicated between experiments. The standard curve may be obtained by using folic acid at levels of 0.0, 0.1, 0.2, 0.3, 0.4 and 0.5 ng per assay tube (5 ml).

Results and Discussion

Computational Results and Discussion

To scout the complete yield horizon, we first found all the solutions that can satisfy the metabolic constraints for the growth of *E. coli* in glucose minimum medium *while* maximizing folic acid flux. We found 32 different flux distributions. The highest carbon yield predicted for folic acid [mol C in folic acid/mol C from glucose] is 0.087. Certain aspects of the flux solutions will be highlighted further here because of their potential implementation via metabolic engineering and a tractable relationship exists to what is known about folic acid synthesis. One aspect is the role of PYK activity (i.e. Figure 1; r_{18} flux). As noted earlier, reducing the activity of PYK could elevate at least one folic acid precursor, PEP, as well as divert carbon normally lost to acids (*11, 12*) into other more useful metabolic products.

Phase plane projections are very useful to present the flux data. It projects the solution space (a convex polytope) into a two dimensional region of the stoichiometrically feasible values of two fluxes. One such phase plane for folic acid production by wild-type *E. coli* is shown in Figure 2, where PYK-catalyzed flux is the independent variable. The triangle gives the solutions space of folic acid production rate and PYK flux. Any point inside the triangle is a feasible solution. The maximum folic acid synthesis rate increases linearly as the PYK-catalyzed flux decreases to zero. This suggests that a mutant deficient in PYK activity may exhibit elevated folic acid production.

Because PEP is an important folic acid precursor, another potential strategy for elevating folic acid synthesis is to replace the phospho-transferase system (PTS) with a glucose permease. In *E. coli*, one mole of PEP is converted to pyruvate when one mole of glucose is transported into the cell via PTS. Replacing the PTS by glucose permease, which transports one mole of glucose at cost of one mole of ATP, could thus reduce PEP use and increase its intracellular concentration.

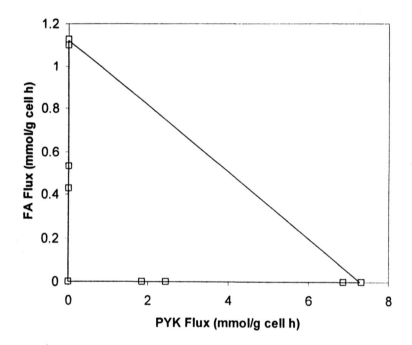

Figure 2. *Phenotype phase plane of wild-type E. coli. PYK-catalyzed flux is constrained from 0 to 20 mmol g^{-1} h^{-1}. The open squares are the projection of extreme points of soluction space.*

A comparison of the PTS and PYK activity-modulating strategies and abbreviated pathways are shown in Figure 3. Based on the wild-type phase plane (Figure 2) and the stoichiometry, models of PYK- and PTS-deficient strains were constructed. The phase planes for the PYK- and PTS-deficient strains are shown and compared to the wild-type in Table 1. These planes illustrate the stoichiometrically feasible values of folic acid production rate and glucose uptake rate. It shows the maximum folic acid production rate for different values of glucose uptake rate.

The phase plane for the PYK-deficient strain (Table 1) is simply a subspace of the wild-type because one additional constraint was added (i.e. Figure 1, $r_{18} = 0$). Therefore, the maximum production potential of folic acid for the wild-type cannot be surpassed by a PYK mutant. However, the solution shows a contracted glucose uptake window within which, folic acid synthesis potential is high. Moreover, the phase plane for the single PYK mutation provides a bench mark for comparison to the folic acid production that may be achieved by the PTS deletion.

The PTS-deficient mutant is predicted (Table 1) to have greater folic acid production potential compared to the wild-type (and the PYK-deficient mutant). The maximum production rate is predicted to equal 1.81 mmol g^{-1} h^{-1}. Moreover, this high potential production rate could also be achieved with a lower to comparable glucose uptake rate (9 mmol g^{-1} h^{-1}). This outcome indicates that the potential carbon yields are higher than those associated with the wild-type or PYK-deficient strains. Indeed, based on the fluxes (not shown), the maximal potential yield for the PTS-deficient strain is 0.211 mol folic acid/mol glucose, which is more than double the wild-type's or PYK-deficient mutant's yield potential.

The results show that PTS mutation might be an effective strategy for increasing folic acid production. However, a recent study (13) reported that PTS-deficient strains can exhibit very slow growth rates. Slow growth could potentially limit the utility of the PTS-deletion strategy for folic acid production especially if the product was labile and regulation is such that product synthesis is growth-related.

Experimental Results and Discussion

Based on desirable growth properties (12) and stoichiometric potential, we have first focused on examining the utility of using a PYK-deficient mutant of *E. coli* for producing folic acid. Figure 4 shows that the growth rates of the two strains were similar. Additionally, folic acid accumulation in the medium appears to parallel growth. However, the folic acid production by the PYK-deficient mutant (0.27mg/L at the end of growth) was significantly higher as

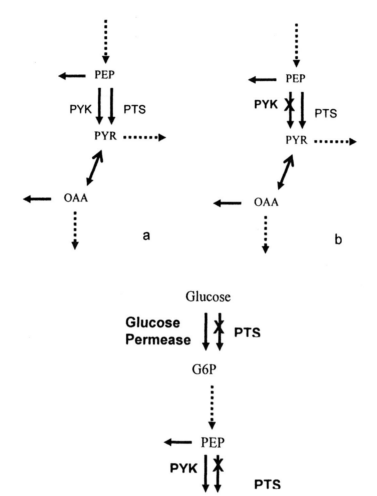

Figure 3. Schematic presentation of three metabolic engineering strategies to
increase folic acid production and representative phenotypephase planes. The
scenarios are (a) E. coli wild-type, (b) a mutation that deletes PYK activity, and
(c) a mutation that replaces PTS with a glucose permerase.

Table 1. Phase Planes of Different *E. coli* Strains

Strain	*Phase Planes of Folic Acid Production vs. Glucose Consumption*
Wild Type	
PYK-Deficient Mutant	
PTS-Deficient Mutant	

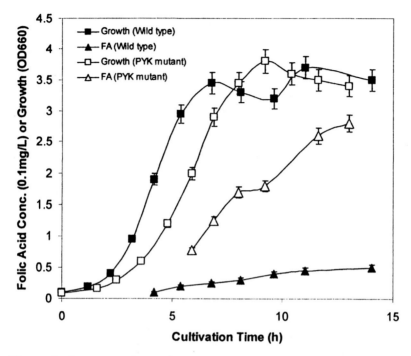

Figure 4. *Experimental measurements of how folic acid production in an E. coli pyk- mutant (PB25) compares to the wild-type (JM101). Because the cell concentrations were comparable at each time point, the raw absolute folic acid activities measured by the bioassay are shown and can be directly compared.*

compared to the wild-type strain (0.05mg/L at the end of growth). Thus, PYK-deficient mutants show promise as a starting point for enhancing the folic acid production by bacteria. While The folic acid production in PYK mutant is substantially higher than the wild-type *E. coli*, the production level is still too low to constitute a viable strategy for commercial production of folic acid. Further metabolic enginnering work is required for design of efficient folic acid producing *E. coli* strains. Our other initial work has also suggested (not shown) that after accumulating in the medium, folic acid is consumed by nutrient-starved cultures presumably for the glutamate content. Future work will be devoted to mapping out the product formation and degradation kinetics in more detail in order to further maximize the product yield. Additionally, using what has been learned in *E. coli* to metabolically engineer *Bacillus subtilis* to produce folic acid with high yield is envisioned. In contrast to *E. coli*, *B. subtilis* is generally regarded as "safe;" hence, *B. subtilis* may be prove to be a more commercially acceptable platform for nutraceutical production.

218

Acknowledgment

MMD and MMA acknowledge the partial support of this work from NSF Grant BES-0118961 and BES-0224603.

References

1. Fell, D. A., & Small, J.R. Fat synthesis in adipose tissue: an examination of stoichiometric constraints *Biochemical J.*, **1986**, 238, 781-786.
2. Majewski RA, Domach MM. Simple constrained optimization view of acetate overflow in E. coli. *Biotechnol. Prog.* 1990. 35, 732-738.
3. Varma A, Palsson, B.O. Stoichiometric flux balance models quantitatively predict growth and metabolic by-product secretion in wild type Escherichia coli W3110. *Appl. Env. Microbiol.* 1994. 60, 3724-3731.
4. Lee S, Phalakornkule C, Ataai MM, Domach MM, Grossmann IE. Recursive MILP Model for Finding All the Alternate Optima in LP Models. *Comput. Chem. Eng.* 2000. 24, 711-716
5. Zhu T, Domach MM, Ataai MM. Software development of convex analysis of stoichiometric model and NMR simulation. *Metabolic Engineering.* Submitted.
6. Mandelstam J, McQuillen K, and Dawes I. "Biochemistry of Bacterial Growth", Halsted Press, New York. **1982.**
7. Ingraham JL, Maaloe O, Neidhardt FC. "Growth of the Bacterial Cell", Sinauer Associates, Sunderland, MA. **1983.**
8. Lee J., Goel A., Ataai M.M. and Domach M.M. Supply-side analysis of growth of *B. subtilis* on glucose-citrate medium: feasible network alternatives and yield optimality. *Appl. Env. Microbiol.* **1997**, 63, 710-718.
9. Ponce E., Flores N., Martinez A., Valle F., and Bolivar F. Cloning of the two pyruvate kinase isoenzyme structural genes from *Escherichia coli*: the relative roles of these enzymes in pyruvate biosynthesis. *J. Bact.* **1995**, 177, 5719-22.
10. Maniatias, T., Fritsch, E. F., Sambrook, J. Molecular cloning: a laboratory manual; Cold Spring Harbor Press: New York, **1982.**
11. Fry B., Zhu T., Phalakornkule C., Koepsel R., Domach M. M., Ataai M. M. 2000. Characterization of Cell Growth, Acid Production In Pyruvate Kinase Mutant of *B. subtilis*. *Appl. Env. Microbiol.* **66**:4045-4049.

12. Zhu T., Phalakornkule C., Koepsel R. R., Domach M. M., Ataai M. M., 2001. Cell Growth and By-Product Formation in a Pyruvate Kinase Mutant of *E. coli. Biotechnol. Prog.* **17**: 624-628.

13. Flores, S., Gosset, G., Flores, N., Graaf, A. A. and Bolívar F.. Analysis of Carbon Metabolism in Escherichia coli Strains with an Inactive Phosphotransferase System by 13C Labeling and NMR Spectroscopy. *Metabolic Engineering.* **2002**, 4,124–137.

Chapter 14

Production of 5-Aminolevulinic Acid and Vitamin B$_{12}$ Using Metabolic Engineering of *Propionibacterium freudenreichii*

**Pornpimon Kiatpapan[1], Nitjakarn Kanamnuay[2],
Boonsri Jongserijit[2], Yong Zhe Piao[3], Mitsuo Yamashita[3],
and Yoshikatsu Murooka[3,*]**

[1]Biochemistry Unit, School of Science, Rangsit University,
Patumthani 12000, Thailand
[2]Department of Biology, Faculty of Science, Silpakorn University, Nakorn,
Pathom 73000, Thailand
[3]Department of Biotechnology, Graduate School of Engineering, Osaka
University, Yamada-oka, Suita, Osaka 565–0871, Japan

Metabolic engineering of *Propionibacterium* sp. by expression of genes involved in vitamin B$_{12}$ biosynthesis pathway is reviewed. The *Rhodobacter hemA* encoding 5-aminolevulinic acid (ALA) synthase and *Propionibacterium hemB* and *cobA* encoding ALA dehydratase and uroporphyrinogen III methyltransferase, respectively, were expressed in *P. freudenreichii* subsp. *shermanii* IFO12426 using vector pPK705 and *Propionibacterium* promoters. Productions of ALA and vitamin B$_{12}$ were studied in strain IFO12426 carrying plasmid-contained *hemA*, *hemB*, or *cobA*. Production of ALA and vitamin B$_{12}$ were 4- and 2-fold enhanced, respectively, in recombinant *Propionibacterium*.

Vitamin B_{12} is one of the most complicated non-polymeric molecules biosynthesized in cells and is present in organisms belonging to the three kingdoms, eubacteria, archaebacteria, and eukaryotes. Vitamin B_{12}, including its coenzyme form, either deoxyadenosylcobalamin or methylcobalamin, is an important coenzyme that is used as a cofactor in a number of enzyme-catalyzed rearrangement and methylation reactions and is able to donate a large chemical potential to its protein counterpart by utilizing the central cobalt atom to perform chemical reactions. Exclusively microorganisms synthesize vitamin B_{12}. Animals requiring this vitamin meet this need by food intake or by absorption of the vitamin produced by animal intestinal microorganisms. Humans, unable to absorb vitamin B_{12} produced in the large intestine, are dependent on food intake for this vitamin. In humans, a deficiency of vitamin B_{12} causes pernicious anemia and peripheral neurological disorder. The commercial production of vitamin B_{12} is currently carried out by microbial fermentation using strains of *Propionibacterium* or *Pseudomonas*. Several review articles for vitamin B_{12} synthesis have been published (1-6). However, no scientific work for vitamin B_{12} production with genetically engineered strains has been reported except some patents (7, 8). Here, we will review our recent works of production of 5-aminolevulinic acid (ALA) and vitamin B_{12} by genetically engineered *Propionibacterium freudenreichii*.

In all organisms, the synthesis of hemes, cobalamin, chlorophylls and vitamin B_{12} starts with the formation of 5-aminolevulinic acid (ALA) and proceeds through the formation of porphobilinogen (PBG) and uroporphobilinogen (UPB). ALA is a useful metabolite since it is utilized as a biodegradable herbicide, insecticide, for photodynamic cancer therapy, or plant growth hormone (9). ALA is synthesized by either of two pathways, the C_4 (Shemin pathway) and the C_5 pathways (Fig. 1). In aerobic and aerotolerant microorganisms studied thus far, ALA is formed by the condensation of glycine and succinyl CoA (Shemin pathway) catalyzed by ALA synthase. The C5 pathway has been reported in archeabacteria, anaerobic bacteria, and the facultative anaerobe *Propionibcterium shermanii* (10). ALA is synthesized from glutamate by a series of reactions, which include the activation of glutamate by its ligation to tRNA, reduction of the activated glutamate by an NADPH dependent reductase to yield glutamate1-semialdehyde (GSA) and transamination of GSA by a GSA 2,1-aminotransferase to form ALA. Genes involved in ALA and vitamin B_{12} biosynthesis have been studied in *Propinibacterium freudenreichii* (11). The *hemL* gene that encodes GSA 2,1-aminotransferase was identified by complementation of an ALA-deficient mutant (*hemL*) of *E. coli* (10). The *hemB* gene encoding PBG synthase (ALA dehydratase) in *P. freudenreichii* was cloned by complementation to the *hemB* mutant of *E. coli* (12). The *cobA* gene encoding uroporphyrinogen III methyltransferase was identified and overexpressed in *E. coli* (13). To date, no *hemA* gene encoding ALA synthase in propionibacteria has been reported.

Overexpression of genes involved in the biosynthetic pathway of vitamin B_{12} should facilitate the production of ALA and vitamin B_{12}. Success in genetic manipulation and gene expression of *Propionibacterium* opens the possibility of genetic study and molecular breeding of propionibacteria (*14*).

Development of an expression vector in propionibacteria

To construct an expression vector for heterologous genes in propionibacteria, an appropriate vector and promoter are required. Recently, shuttle vectors for shuttling between *Propionibacterium* and *Escherichia coli* were constructed using propionibacterial replicon and an appropriate selection marker. Vector pPK705 was constructed by us (*15*) from endogenous plasmid pRGO1 from *P. acidipropionici*, pUC18 (*16*) and the *Streptomyces hygromycin B* resistant gene (*17*). Another vector pBRESP36A was constructed using the p545 replicon, pBR322 and the erythromycin resistant gene from *Saccharopolyspora erythraea* (*18*). The vector pPK705 could transform several species of *Propionibacterium* at high efficiency using the vector prepared from *Propionibacterium* cells to overcome a high restriction–modification system in propionibacteria (*15*). The vector pBRESP36A transformed only *P. freudenreichii* at high efficiency (*18*). The vector pPK705 was developed to express heterologous genes in propionibacteria. Several native promoters from *P. freudenreichii* subsp. *shermanii* IFO12424 were isolated using a pCVE1 promoter probe vector containing the *Streptomyces* cholesterol oxidase gene (*choA*) as a reporter gene (*19*), since promoters from gram-negative bacteria, such as *lac* or *tac*, could not drive any gene in propionibacteria. These propionibacterial promoters expressed the *Streptomyces choA* gene in *E. coli* (*20*). Two of four strong promoters, which expressed the *Streptomyces choA* gene in *P. freudenreichii* subsp. *shermanii*, were selected to construct an expression vector for use with propionibacteria. Preferably, the gene to be expressed should be a gram-positive bacterium of high GC content. Thus, the expression vectors, P1N and P4N, were constructed using propionibacterial promoters, P1 and P4, three frame stop codons from pCVE1 and the ribosome-binding site (SD sequence) and the start codon of *choA* which contains the synthesized *NcoI* restriction enzyme site. An interesting gene to be expressed could be inserted at the *NcoI* site fused to the start codon of *choA*. Transcription of a heterologous gene in propionibacteria occurs through the SD sequence and start codon of *Streptomyces choA* (*21*).

Production of 5-aminolevulinic acid (ALA)

The expression of the *Streptomyces choA* gene in *P. freudenrichii* subsp. *shermanii* under the control of propionibacterial promoter led to construction of

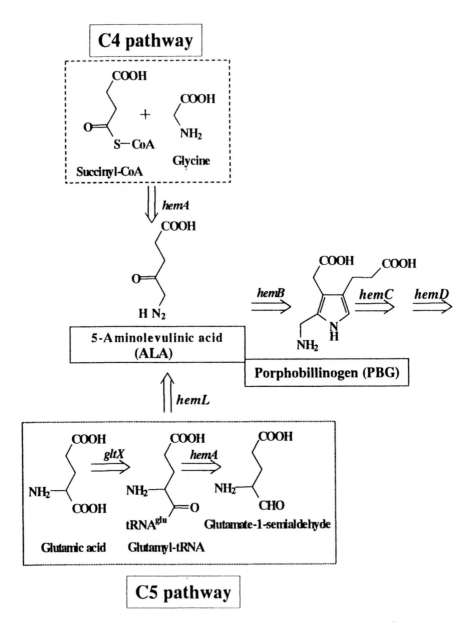

Figure 1. Biosynthesis pathway of tetrapyrrole, vitamin B$_{12}$, heme, and

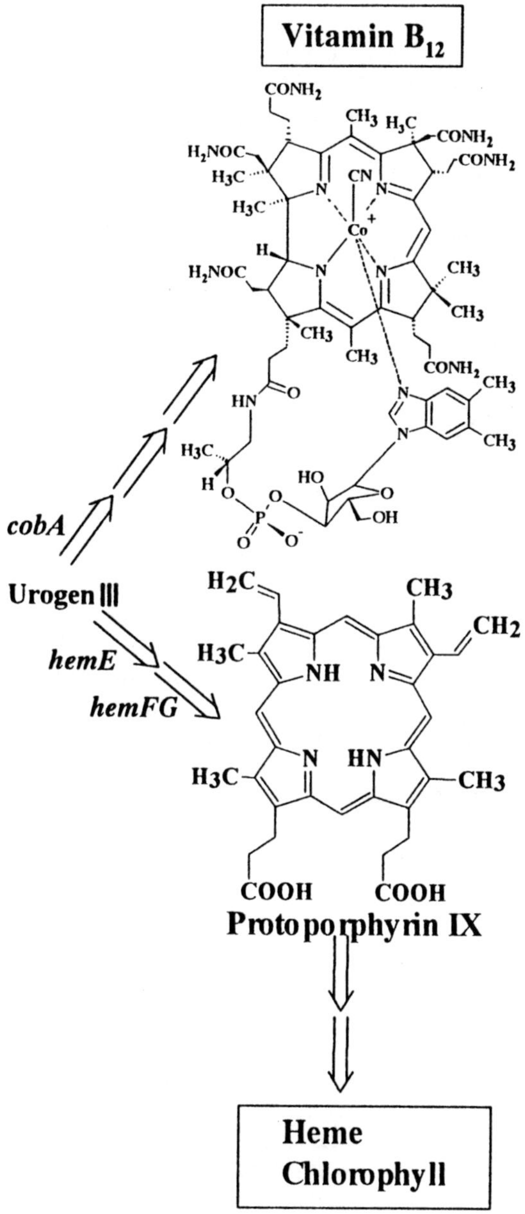

Vitamin B₁₂

cobA

Urogen III

hemE

hemFG

Protoporphyrin IX

**Heme
Chlorophyll**

chlorophyll via ALA.

an expression vector for genes involved in the vitamin B_{12} biosynthesis pathway. Production of ALA by microorganisms has been reported in *Clostridium thermoaceticum (22)*, methanogenes *(23)*, *Chlorella* spp. *(24-26)*, and photosynthetic bacteria *(24, 27-34)*. ALA production by photosynthetic bacteria, however, requires light illumination and has been found to be sensitive to aeration. A crude extract from recombinant *E. coli*, which contains the *hemA* encoding ALA synthase, has been shown to synthesize ALA with high yield but requires a large amount of expensive ATP *(27)*. These problems have provided significant barriers to the production of ALA on an industrial scale. To address this, Nishikawa et. al. *(34)* have introduced a metabolic engineering approach based on sequential mutagenesis of *Rhodobacter spheroides* for the production of ALA in the absence of light illumination. In this way, industrial production of ALA has been successful *(35)*.

To overproduce ALA in propionibacteria by metabolic engineering, the ALA synthase gene (*hemA*) from *R. sphaeroides* was recombined in the expression vectors in *P. freudenreichii* subsp. *shermanii* IFO12426 *(21)*. The *hemA* gene from *R. sphaeroides* driven by *Propionibacterium* promoters was expressed, and ALA was accumulated in strain IFO12426 carrying pKHEM01 or pKHEM04 as compared to the strain carrying pPK705 (Table I). ALA was accumulated 0.94 mM in strain IFO12426 carrying pKHEM04. No significant increase of ALA production by addition of glycine and succinate, which are known as precursors of ALA in the C_4 pathway, nor by addtion of levulinic acid, an inhibitor of PBG synthase. The P4 promoter was better than the P1 promoter in *P. freudenreichii*. Plasmid pKHEM06 was constructed by inserting the *hemB* gene and the *cobA* gene from *P. freudenreichii* IFO12426 immediately downstream of the *hemA* gene in plasmid pKHEM04. The accumulation of PBG was increased in the strain carrying pKHEM06, whereas the accumulation of ALA was decreased compared to the strain carrying pKHEM04. The increase of PBG production may be due to the expression of the *hemB* gene under the control of the P4 promoter.

Production of Vitamin B12

Since the chemical synthesis of vitamin B_{12} involves a highly complicated, 70 step synthesis, vitamin B_{12} has been produced at industrial scale by fermentation using selective strain improvement by mutation followed by optimization of the fermentation process. Random mutagenesis is frequently used to improve production of vitamin B_{12} in several B_{12} producing species including *Pseudomonas* sp. and *Propionibacterium* sp. By using random mutagenesis combined with methods of genetic engineering, a production of 300 mg /l was achieved in *Pseudomonas denitrificans* *(36)*. According to the European patent *(8)*, eight genes in the *cobF-cobM* operon and *cobA* and *cobE*

TABLE I. Production of ALA, PBG and Vitamin B$_{12}$ by *P. freudenreichii* subsp. *shermanii* IFO12426 Carrying Various Plasmids

Plasmid (promoter-gene)	ALA^a (mM)	PBG^a (mM)	Tetrapyrroles^b (mg/l)	Vitamin B$_{12}$^{b} (mg/l)
pPK705 (*hemA*)	0.53	0.03	0.10	0.34
pKHEM01 (P1-*hemA$^+$*)	4.41	0.17	0.60	0.58
pKHEM04 (P4-*hemA$^+$*)	10.9	0.31	1.2	0.69
pKHEM06 (P4-*hemA$^+$ hemB$^+$ cobA$^+$*)	10.2	1.12	1.4	1.2

^{a}Cells were grown in GYT medium (1.0% glucose, 0.5% yeast extract, 1.0% trypticase soy broth) with 0.5% (NH$_4$)$_2$SO$_4$, 10 mg/l CoCl$_2$6H$_2$O and 125 µg/ml hygromycin B at 32 °C for 48 h after which cells were grown for another 24h.

^{b}Cells were grown in GYT medium with 125 µg/ml hygromycin B at 32 °C for 72 h. Tetrapyrroles and vitamin B$_{12}$ were analyzed by HPLC analysis.

P1- and P4-*hemA$^+$ hemB$^+$ cobA$^+$* in parentheses indicate that the *hemA*, *hemB* or *cobA* genes were under the control of the P1 and P4 promoters, respectively.

genes were amplified in *P. denitrificans*. The amplification of the *cobF-cobM* gene cluster and *cobAE* genes enhanced about 30% and 20% cobalamin production, respectively. The vitamin B_{12} production by sequential mutagenesis of *P. denitrificans* was increased about 100-fold than the producer strain (8). These results suggest that release of feedback regulation is more effective than the gene amplification on the overproduction of vitamin B_{12} by *Pseudomonas*. Biochemistry of enzyme reactions involved in biosynthesis of vitamin B_{12} were extensibly worked in *P. denitrificans* by Rhone-Pulence Rorer's group. They clarified the conversion of precorrin-6x from ALA and trimetylisobacterocholin (37), precorrin-6x octamethyl ester synthesis (38), amidation of 5'-deoxy-5'-adenosyl-cobyrinic acid a,c-diamide (39), and precorrin-6y to precorrin-8x (40) by cell-free protein from *P. denitrificans*. They also demonstrated that precorrin-3 is converted into a further trimethylated intermediate, precorrin-3b (41), with cell-free protein prepared from a series of recombinant strains of *P. denitrificans*.

Production of vitamin B_{12} using *Propionibacterium* spp. has been thought to be more acceptable to consumers since strains of dairy propionibacteria are commonly used in dairy food industries. The *cobA* gene encoding uroporphyrinogen (urogen) III methyltransferase, a key enzyme in the biosynthetic pathways of vitamin B_{12} in *P. freudenreichii* was studied and expressed in *E. coli* (13). Expression of the *cobA* gene was performed through the *Propionibacterium* strain and proved to increase the production of vitamin B_{12} (5). Recently, the *cobA* gene from *P. freudenreichii* was added to *hemA* from *R. spheroides* and *hemB* from *P. freudenreichii* and gave rise to plasmid pKHEM06. The production of vitamin B_{12} by expression of the *hemA*, *hemB* and *cobA* genes under the control of various promoters in *P. freudenreichii* subsp. *shermanii* was studied (Table I). In all recombinant propionibacteria, the production of vitamin B_{12} was increased. The production of vitamin B_{12} by the recombinant strain carrying pKHEM04 is higer than that carrying pKHEM01, which may be due to the increased production of ALA by pKHEM04. Although the production of PBG was about 4 fold higher in the strain carrying pKHEM06 than that in the strain carrying pKHEM04, about 2-fold increase of vitamin B_{12} production was observed. The observation of reddish color in the culture of cells carrying pKHEM06 suggests the production of other tetrapyrrole compounds by these recombinant cells. We found that significant amounts of several tetrapyrrole compounds were accumulated in the cells carrying pKHEM01, pKHEM04 and pKHEM06 (Table 1). Sorts of tetrapyrrole compounds accumulated in strains carrying plasmids were different (Piao et al., unpublished results). These results suggest that blocking of HemE or HemFG biosynthesis could change the flow from heme pathway to vitamin B_{12} pathway (Fig. 1). This flow change may result in facilitate the production of vitamin B_{12} from the precursors. The time course of vitamin B_{12} production by the recombinant strain carrying pKHEM06 was tested. Production of vitamin B_{12} was increased linearly over 3 days cultivation coordinated with cell growth (Fig. 2).

The bioprocess of vitamin B_{12} production using *Propionibacterium* strains may be divided into two stages (42). The bacteria required anaerobic

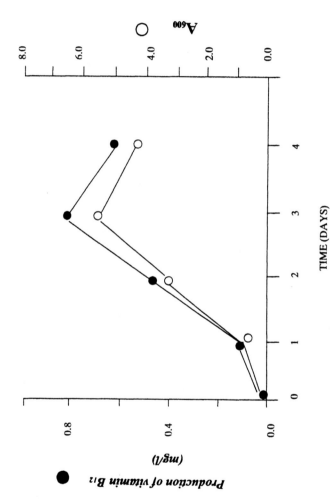

TIME (DAYS)

Figure 2. Time course of production of vitamin B$_{12}$ in recombinant P. freudenreichii subsp. shermanii IFO12426 carrying pKHEM06. The cells were grown in GYT medium (1.0% glucose, 0.5% yeast extract, 1.0% trypticase soy broth) with 0.5% (NH4)2SO4, 10 mg/l CoCl2/6H2O and 125 µg/ml hygromycin B at 32 for 72h. closed and open circles indicateVitamin B12 concentration and A$_{600}$, respectively.

fermentation conditions in the first three days to produce the vitamin B_{12} precursor cobamide, a vitamin B_{12} intermediate missing the 5,6-dimethylbenzimidazole (DMBI). Subsequently vitamin B_{12} formation was completed by gentle aeration of the whole culture for 1-3 days, allowing the bacteria to undertake the oxygen-dependent synthesis of DMBI and to link it to cobamide. Furthermore, it is crucial to neutralize the accumulated propionic acid during the whole fermentation process, in order to maintain the production culture at pH 7, since the formation of propionic acid amounts to 10% volume of the fermentation volume (*43*). Ye et al. (*44*) studied performance of *P. freudenreichii* in anaerobic, aerobic and periodic fermentations. Vitamin B_{12} was synthesized in the anaerobic cycle and ceased the production in the aerobic cycle. The B_{12} production increased by about 2-fold as compared with that attained in the anaerobic fermentation.

In contrast to the *Propionibacterium* fermentation process, the production of vitamin B_{12} using *Pseudomonas* sp. paralleled oxygen-dependent growth with high vitamin B_{12} production rates. The culture was aerated during the whole fermentation process of about 2-3 days at 30 °C with pH levels of 6-7 (*43, 45*).

Independent of the employed production strains and culture condition, it seems to be necessary to add some essential compound to the medium for efficient vitamin B_{12} biosynthesis (*46*). The addition of cobalt ions and DMBI are frequently described. Thus, the fermentation process of vitamin B_{12} production is quite complicated, so that further study on metabolic engineering by *P. freudenreichii* subsp. *shermanii* will give us more information for high yields of vitamin B_{12} and other tetrapyrrole compounds.

References

1. Battersby, A. R. *Science*, **1994**, *264*, 1551-1557.
2. Santander, P. J; Roessner, C. A.; Stolowich, N. J.; Holdeman, M. T.; Scott, A. I. *Chemisry & Biology*, **1997**, *4*, 659-666.
3. Raux, E.; Schubert, H. L.; Rope, J. M.; Wilson, K. S.; Warren, M. J. *Bioorganic Chem.*, **1999**, *27*, 100-118.
4. Scott, A. I. *The Chemical Record*, **2001**, *1*, 212-227.
5. Martens, J. –H.; Barg, H.; Warren, M. J.; Jahn, D. *Appl. Microbiol. Biotechnol.* **2002**, *58*, 275-285.
6. Warren, M. J.; Raux, E.; Schubert, H. L.; Escalante-Semerena, J. C. *Nat. Prod. Rep.* **2002**, *19*, 390-412.
7. Blanche, F.; Cameron, B.; Crouzet, J.; Debussche, L.; Thibaut, D.; *World Patent* 97/43421, **1997.**
8. Blanche, F.; Cameron, B.; Crouzet, J.; Debussche, L.; Levy-Schil, S,; Thibaut, D.; *Eur. Patent* 0516647, **1998.**
9. Nishikawa, S.; Murooka, Y. *Biotechnology and Genetic Engineering Reviews*, Harding S. E. Ed.; Andover, UK, 2001, vol. 18, pp. 149-170.

10. Murakami, K.; Hashimoto, Y.; Murooka, Y. *Propionibactertium freudenreichii*. *Appl. Environ. Microbiol*. **1993**, *59*, 347-350.
11. Hashimoto, Y.; Yamashita, M.; Murooka, Y. *Appl. Microbiol. Biotechnol*., **1997**, *47*, 385-392.
12. Hashimoto, Y.; Yamashita, M.; Ono, H.; Murooka, Y. *J. Ferment. Bioeng*. **1996**, *82*, 93-100.
13. Sattler, I.; Roessuner, C. A.; Stolowich, N. J.; Hardin, S. H.; Harris-Haller, L. W; Yokubaitis, N. T.; Murooka, Y.; Hashimoto, Y.; Scott, A. I. *J. Bacteriol*., **1995**, *177*, 1564-1569.
14. Kiatpapan, P.; Murooka, Y. *J. Biosci. Bioeng*. **2002**, *93*, 1-8.
15. Kiatpapan, P.; Hashimoto, Y.; Nakamura, H.; Piao Y. Z.; Yamashita, M.; Ono, H.; Murooka, Y. *Appl. Environ. Microbiol*., **2000**, *66*, 4688-4695.
16. Yanisch-Perron, C.; Vieira, J.; Messing, J. *Gene*, **1985**, *33*, 103-119.
17. Zalacain, M.; Gonzalez, A.; Guerrero, M. C.; Mattaliano, R. J.; Malpartida, F.; Jimenez, A. *J Nucleic Acids Res*., **1986**, *14*, 1565-1581.
18. Jore J. P. M.; Lujik van N.; Luiten, R. G. M.; Werf van der M. J.; Pouwels, P. H. *Appl. Environ. Microbiol*., **2001**, *67*, 499-503.
19. Nomura, N.; Choi, K.-P.; Yamashita, M.; Yamamoto, H.; Murooka, Y. *J. Ferment. Bioeng*., **1995**, *79*, 410-416.
20. Kiatpapan, P.; Yamashita, M.; Kawaraichi, N.; Yasuda, T.; Murooka, Y. *J. Biosci. Bioeng*., **2001**, *92*, 459-465.
21. Kiatpapan, P.; Murooka, Y. *Appl. Microbiol. Biotechnol*., **2001**, *56*, 144-149.
22. Koesnandar, A. S.; Nishio, N.; Nagai, S. *Biotechnol Lett*., **1989**, *11*, 567-572.
23. Lin, D.; Nishio, N.; Nagai, S. *J. Ferment. Bioeng*., **1989**, *68*, 88-91.
24. Sasaki, K.; Watanabe, K.; Tanaka T.; Hotta Y.; Nagai, S. *World Journal of Microbiology Biotechnology*, **1995**, *11*, 361-362.
25. Ano, A.; Funahashi, H.; Nakao, K.; Nishizawa, Y. *J. Biosc. Bioeng*., **1999**, 88, 57-60.
26. Ano, A.; Funahashi, H.; Nakao, K.; Nishizawa, Y. *J. Biosc. Bioeng*. **2000**, *89*, 176-180.
27. Van Der Werf, J. W.; Zeikus, J. G. *Appl. Environ. Microbiol*., **1996**, *62*, 3560-3566.
28. Sasaki, K.; Ikeda, S.; Konishi, T.; Nishizawa, Y.; Hayashi, M. *J. Ferment. Bioeng*., **1989**, *68*, 378-381.
29. Sasaki, K.; Tanaka, T.; Nishizawa, Y.; Hayashi, M. *Appl. Microbiol. Biotechnol*., **1990**, *32*, 727-731.
30. Sasaki, K.; Tanaka, T.; Nagai, S. *Biotechnol. Lett*., **1993**, *15*, 859-864.
31. Tanaka, T.; Watanabe, K.; Hotta, Y.; Lin, D.; Sasaki, K.; Nagai, S. *Biotechnol. Lett*., **1991**, *13*, 589-594.
32. Tanaka, T.; Sasaki, K.; Naparatnaraporn, N.; Nishio, N. *World Journal of Microbiology Biotechnology*, **1994**, *10*, 677-680.

33. Tanaka, T.; Watanabe, K.; Nishikawa, S.; Hotta, Y.; Sasaki, K.; Murooka, Y.; Nagai, S. *Seibutsu Kogaku Kaishi*, **1994**, *72*, 461-467.

34. Nishikawa, S.; Watanabe, K.; Tanaka, T.; Miyachi, N.; Hotta, Y.; Murooka, Y. *J. Biosci. Bioeng.*, **1999**, *87*, 798-804.

35. Kamiyama, H.; Hotta, Y.; Tanaka, T.; Nishikawa, S.; Sasaki, K. *Seibutsu Kogaku Kaishi*, **2000**, *78*, 48-55.

36. Blanche, F.; Cameron, B.; Crouzet, J.; Debussche, L.; Thibaut, D.; Vuyilhorgne, M.; Leeper, F. J.; Batterby, A. R. *Angew. Chem.*, **1995**, *107*, 421-452.

37. Thibaut, D.; Debussche, L.; Blanche, F. *Proc. Natl. Acad. Sci. USA*, **1990**, *87*, 8795-8799.

38. Thibaut, D.; Blanche, L.; Debussche, F.; Leeper, F. J.; Battersby, A. R. *Proc. Natl. Acad. Sci. USA*, **1990**, *87*, 8800-8804.

39. Blanche, F.; Couder, M.; Debussche, L.; Thibaut, B.; Cameron, B.; Crouzet, J. *J. Bacteriol.* **1991**, *173*, 6046-6051.

40. Blanche, F.; Famechon, D.; thibaut, l.; Debussche, B.; Cameron, B.; Crouzet, J. *J. Bacteriol.* **1992**, *174*, 1050-1052

41. Debussche, l.; thibaut, B.; Cameron, B.; Crouzet, J.; Blanche, F. *J. Bacteriol.* **1993**, *175*, 7430-7440.

42. Quesada-Chanto, A.; Afschar, A. S.; Wagner, F., *Appl. Microbiol. Biotechnol.*, **1994**, *41*, 378-383.

43. Eggersdorfer, M., *Vitamins*, In: Ullmann's encyclopedia of industrial chemistry, vol 27A, 5th edn., Elvers B, Hawkinds S (eds), VCH, Weinheim,1996, pp 443-613.

44. Ye, K.; Shijo, M.; Jin, S.; Shimizu, K. *J.Fferment. Bioeng.* **1996**, *82*, 484-491.

45. Scott, J. W., *Vitamin B12*. In: Kirk-Othman encyclopedia of chemical technology, vol 25, 4th edn. Kroschwitz J. I., Howe-Grant, M. (eds), Wiley, New York, 1998; pp 137-217.

46. Bykhovsky, V. Y.; Zaitseva N. I.; Eliseev, A. A. *Appl. Biochem. Microbiol.*, **1998**, *34*, 1-18.

Chapter 15

Metabolic Flux Analysis Based on Isotope Labeling Technique and Metabolic Regulation Analysis with Gene and Protein Expressions

Kazuyuki Shimizu[1,2,*]

[1]Department of Biochemical Engineering and Science, Kyushu Institute of Technology, Iizuka, Fukuoka 820–8502, Japan
[2]Current address: Institute of Advanced Biosciences, Keio University, Tsuruoka, Yamagata 997–0017, Japan
*Telephone: +81–948–29–7817; fax: +81–948–29–7801; email: shimi@bse.kyutech.ac.jp

The recent progress on metabolic systems engineering in particular metabolic flux analysis (MFA) based on the isotopomer distribution obtained using NMR and/or GC-MS was reviewed. After the brief explanation on how to estimate the metabolic flux distribution (MFD), the comparison was made between MFD and gene and protein expressions for cyanobacteria and *Escherichia coli* to investigate the metabolic regulation. The integration of these information was found to be very important to uncover the metabolic regulation of microorganisms in relation to the genetic change and/ or the change in the culture condition.

Introduction

Recent rapid progress in molecular biology unveiled the intricacies and mechanistic details of genetic information and determined the structure of DNA and the nature of the gene code, establishing DNA as the source of heredity containing the blueprints which organisms are built. These scientific revolutionary endeavors have rapidly spawned the development of new technologies and emerging fields of research based around genome sequencing efforts (i.e., functional genomics, structural genomics, proteomics and bioinfomatics) [1].

Once presented with the sequence of a genome, the first step is to identify the location and size of genes and their open reading frames (ORF's). DNA sequence data then need to be translated into functional information, both in terms of the biochemical function of individual genes, as well as their systematic role in the operation of multigenic functions.

Genomic technologies can be broadly divided into two groups, namely those that alter gene structure and deduce information from altered gene function and those that observe the behavior of intact genes [2]. The former methods involve the random or systematic alteration of genes across an organism's genome to obtain functional information. Current genome-altering technologies are distinguished by their strategies for mutagenizing and analyzing cell populations in a genome-wide manner.

For the latter, several methods have been developed to efficiently monitor the behavior of thousands of intact genes. DNA chips are tools that fractionate a heterogeneous DNA mixture into unknown components, while a complementary method, called SAGE (Serial Analysis of Gene Expression), has been important for identifying transcripts not predicted by sequence information alone [3]. DeRishi et al. [4] showed how the metabolic and genetic control of gene expression could be studied on a genome scale using DNA microarray technology. The temporal changes in genetic expression profiles that occur during the diauxic shift in *S. cerevisiae* were observed for every known expressed sequence tag in this genome.

One of the key issues in functional genomics is to relate linear sequence information to nonlinear cellular dynamics. Toward this end, significant scientific effort and resources are directed at mRNA expression monitoring methods and analysis, and at the same time the field of proteomics (the simultaneous analysis of total gene expression at the protein level) represents one of the premiere strategies for understanding the relationship between various expressed genes and gene products [5].

On the other hand, metabolic engineering has been paid much attention during the last decade for the fermentation improvement. Typically, the objective of industrial fermentation is either to increase the rate of a desired product or to reduce the rate of undesired side-products, or to decompose the toxic or undesired substances. Metabolic engineering approach has been recognized to be useful to attain such a goal. Of central importance in metabolic engineering is the notion of cellular metabolism as a network. In other words, an enhanced perspective of metabolism and cellular function can be obtained by considering the participating reactions in their entirety, rather than on an individual basis [6]. The intellectual framework and the potential application of metabolic engineering have been reviewed [7,8], and yet another reviews have been made to recognize the importance of metabolic engineering [9-13]. Several books have also been published recently [14,15].

Here, I review the recent progress on metabolic engineering, in particular, on the metabolic flux analysis (MFA) based on the isotopomer distribution.

It should be noted that it is quite important to investigate the metabolic regulation based on MFA in relation to gene and protein expressions to understand the overall picture. But little is investigated on this subject. Here, I also consider this problem based on our recent research results.

Metabolic Flux Analysis

Quantification of metabolic fluxes is essential to understand the metabolic regulation in relation to the change in culture environment for the process improvement. In particular, MFA is essential in determining the cell physiology. MFA provides a measure of the degree of engagement of various pathways in cellular functions and metabolic processes. In MFA, it is quite important to accurately quantitate the metabolic flux distributions (MFDs) in vivo for investigating the effect of culture environments on metabolic regulation. In the conventional approach, the intracellular metabolic fluxes are calculated using a stoichiometric model for all the major intracellular reactions and by applying mass balances around the intracellular metabolites. As inputs to the calculations, a set of measured fluxes, typically the specific uptake rate of substrate and the specific secretion rate of metabolites etc. are provided [16-18]. Metabolic flux distribution can then be estimated using the following stoichiometric equation:

$$Ar = q \qquad (1)$$

where A is a matrix of stochiometric coefficients. r is a flux vector, and q

is a vector of the specific substrate consumption rate and the specific metabolite excretion rates. The weighted least square solution to Eq.(1) is then obtained for the over-determined system as

$$r = (A^T \Phi^{-1} A)^{-1} A^T \Phi^{-1} q \qquad (2)$$

provided that A is of full rank, where Φ is the measurement noise variance-covariance matrix of the measurement vector q. Several computer programs for calculating r have been developed by several researchers [19,20].

Metabolic Flux Analysis Based on Isotope Distributions

The metabolic flux analysis based on the MFDs computed by the above method may be useful to study the metabolic regulation to some extent. However, the application of the above method is limited to relatively simple case. In more complex metabolic network systems, where the system involves cyclic pathways such as TCA cycle etc. where intermediates reenter the cyclic pathway, or the large number of branching points exist or parallel reaction steps like the various anaplerotic reactions exist in the metabolic network, a detailed analysis cannot be done with the above method. Moreover, the forward and backward directions of reversible reaction steps can never be resolved with the above method. In such cases, the application of metabolite balancing either requires another sets of reactions or forced to be lumped together [16]. These problems led to the development of metabolic flux analysis based on carbon labeling experiments. Isotopic tracer experiments are usually conducted using substrate enriched with ^{14}C, which is radioactive, or ^{13}C, which is not radioactive and stable and detectable by NMR and/or GC-MS. For stable isotope, the specific activity is usually expressed as the fractional enrichment of a specific atom within a molecule. The latter approach has been paid much attention during the past decade.

It should be noted that for the flux calculation, neither enzyme activities nor kinetic information on enzymes are required. Only the mass balance for the metabolites and carbon atoms are considered, while the energy balance is not considered. Several assumptions are made for flux calculation such that

(1) The system of concern must be kept in a well-defined stationary physiological state during the measurement procedure, since the

steady-state mass balance is applied for flux calculation. This does not, however, restrict the application to continuous culture but may be applied to batch and fed-batch culture by assuming pseudo-steady state.

(2) For the metabolic pathway of interest, all relevant stoichiometric equations must be defined. Moreover, the fate of all carbon atoms must be defined. Those can be easily found in the biochemistry textbooks as far as the main metabolic pathways are considered.

(3) No mass effects are assumed to present. Namely, the labeling state of a molecule does not influence the rate of its enzymatic conversion. This may be true as far as the liquid phase reactions are considered. It should, however, be careful that mass effects have been observed in certain situations for small molecules like CO_2 [21,22] as stated in Wiechert and Graaf [23].

It should also be noted that there is no need to preset the directionality for any of the fluxes. In general, a relatively large negative value of standard free energy may justify the assumption of irreversibility, that is, the corresponding reaction step may be considered to be unidirectional, while bi-directional reactions are considered for the other cases.

Several analysis methods for MFA based on the isotope distribution have been proposed, and those may be classified roughly into two such as positional representation approach and the isotopomer representation approach. For a metabolite with n carbon labeled backbone, the positional level represents the concentrations of the labeled carbon in each of the n positions of the carbon backbone [18,23,24,25], while the isotopomer level represents the normalized concentrations of each of the 2^n isotopomers [26-31]. The positional representation is simpler than the isotopomer representation, but the former is not straightforward for the data obtained from NMR and GC-MS measurements.

It should be noted that the labeling patterns of the intracellular metabolites are difficult to measure due to the small pool sizes of these metabolites. However, since the amino acids reflect the labeling patterns of a number of important central metabolites through their precursors from the central metabolism and relatively abundant, the labeling patterns of amino acids have been used for elucidation of labeling patterns in the central metabolism [32]. Thus ^{13}C-NMR and GC-MS have been employed to identify indirectly the labeling patterns of the intracellular metabolites. In the isotope labeling experiments, the mixture of 90% of unlabeled or naturally labeled carbon source such as glucose and 10% of uniformly labeled carbon source [U-^{13}C] and/or the first carbon labeled carbon source [1-^{13}C] is often employed, but it should be noted that there are many other possible combinations, and this will affect the relaibility of the computed fluxes from the statistical point of view[33]. In the analysis of carbon labeling patterns of amino acids, it is also necessary to make a correction for the contribution of labeling arising naturally labeled species of ^{15}N, ^{17}O, ^{18}O, ^{13}C, ^{2}H etc. This correction may be carried out using matrix-based methods as given by Lee et al. [34,35] and Wittman and Heinzle [36].

For better description of the bidirectional reactions, they may be expressed in terms of their net flux v^{net} and their exchange flux v^{exch} that quantitates the amount of flux common to forward and backward fluxes as shown by Eq.(19) below. Nonlinear mapping of the exchange fluxes v^{exch} to exchange coefficients $v^{exch[0,1]}$ [23,28] as given below in Eq.(4) may be introduced to overcome the numerical problems arising from very large parameter values. Here, β is a constant on the order of magnitude of the net fluxes.

$$v^{exch} = \min(\vec{v}, \bar{v}) \tag{3}$$

$$v^{exch[0,1]} = \frac{v^{exch}}{\beta + v^{exch}} \tag{4}$$

The strategy of flux determination in a complex metabolic network is now explained. Similar to the conventional metabolite balancing, balances are taken around all isotopomers of a particular metabolite. Then the mathematical framework relating intracellular fluxes and isotopomer measurements is developed. The intracellular fluxes can thus be estimated by a nonlinear least squares fitting procedure. One of the procedures to solve this problem may be as follows:

(1) The vector of free fluxes (v^{net}_{free}, $v^{exch[0,1]}$) is given with an arbitrary

value. The linear constraints on the net fluxes provided by the stoichiometry of metabolic networks are not usually sufficient for a complete determination of all net fluxes. Therefore, in order to fix the remaining degrees of freedom, some net fluxes must be identified that enable the metabolite balance equations to be solved. These fluxes, together with all the exchange coefficients that can not be resolved by metabolite balancing, must be taken as the free fluxes to be optimized.

(2) All the net fluxes are determined by using the stoichiometric equations for key intracellular metabolites and the measured extracellular flux data.

(3) The vector (v^{net}, $v^{exch[0,1]}$) is transformed into the vector ($\bar{v}, \tilde{v}$) by using Eqs.(3) and (4).

(4) The set of isotopomer balance equations is solved iteratively via computer to obtain the isotopomer distribution of each metabolite. In the isotopomer balance equations, the sum of incoming isotopomer fluxes is set equal to the sum of isotopomer fluxes out of the pool. The flux of isotopomer into a metabolite is the sum of the substrate isotopomers that are required to produce the individual product isotopomers in biochemical reactions, weighted by the corresponding reaction rate ($\bar{v}, \tilde{v}$) [25,26]. The isotopomer fractions of substrates should be initialized. Then each of the isotopomer balance equations is solved sequentially for the isotopomer distribution vector of products, and the above procedure is repeated until convergence is achieved.

(5) The calculated isotopomer distributions are transformed into multiplet intensities and mass isotopomer distributions.

(6) Steps 1-5 are incorporated into an optimization program to find the optimal free fluxes that generate the estimated isotopomer measurement data and extracellular fluxes fitted best to the experimental results. The optimal estimation for free net fluxes and exchange coefficients is obtained by minimizing the sum of squares of the deviations between estimated values and measured data. The objective function to be minimized may be defined as

$$F(v) = \sum_{i=1}^{M} (\frac{W_i - E_i(v)}{\delta_i})^2 + \sum_{j=1}^{N} (\frac{Y_j - v_k}{\delta_j})^2 \tag{5}$$

where v is the vector of free fluxes to be optimized in the program, and W_i are the M individual isotopomer measurement data and E_i their corresponding estimated values computed by the assumed values of v. Y is a vector containing the measured data of N extracellular fluxes. v_k is the element of v, which corresponds to the extracellular flux measurement Y_j. δ_i is the absolute measurement error[37].

We recently developed a new optimization method for the above optimization by combining local search method with global search method such as genetic algorithm [38].

Marx et al. [18] combined the information of ^{1}H-detected ^{13}C NMR spectroscopy to follow individual carbons with carbon balances for cultivation of lysine-producing strain of *C. glutamicum*. The result shows that the flux through pentose phosphate pathway is 66.4 % (relative to the glucose input flux of 1.49 mmol/g dry weight h), that the entry into TCA cycle 62.2 %, and the contribution of the succinylase pathway of lysine synthesis 13.7 %.

For the systems having cyclic pathways, Klapa et al. [39] presented a mathematical model to analyze isotopomer distributions of TCA cycle intermediates following the administration of ^{13}C (or ^{14}C) labled substances. Such theory provides the basis to analyze ^{13}C NMR spectra and molecular weight distributions of metabolites. This method was applied to the analysis of several cases of biological significance [40].

Wiechert and his co-workers developed a model for positional carbon labeling systems, which was further extended to general isotopomer labeling systems with statistical treatment [23,24,30]. Mollney et al. [29] compared the different measurement techniques such as ^{1}H NMR, ^{13}C NMR and mass spectrometry (MS) as well as two-dimensional ^{1}H-^{13}C NMR techniques to characterize in more detail with respect to the formulation of measurement equations. Based on these measurement equations, a statistically optimal flux estimator was established. Having implemented these tools, different kinds of labeling experiments were compared using statistical quality measures.

Although many applications have been reported for the flux analysis using NMR, GC-MS is also quite useful and has several advantages over NMR such that GC-MS requires relatively small amount of samples as compared with NMR, and GC-MS gives rapid analysis etc. In GC-MS, the compounds are separated by the gas chromatography, and the mass spectrometry step analyzes the labeling patterns of the compounds as they elute. The mass spectrum of a compound usually contains ions that are produced by fragmentation of the molecular ion (i.e., the ionized intact molecule). These fragments contain different subsets of the original carbon skelton, and the mass isotopomer distributions of these fragments contain information that can, in addition to the information from the molecular ion, be used for analyzing the labeling pattern of the metabolites [27]. Christiensen and Nielsen [41] quantified the intracellular fluxes of *Penicillium chrisogenum* using GC-MS. They found that glycine was synthesized not only by serine hydroxymethyltransferase, but also by threonine aldorase. The formation of cytosolic acetyl-CoA was also found to be synthesized both via the citrate lyase-catalysed reaction and by degradation of the penicillin side-chain precursor, phenoxyacetic acid.

We also used GC-MS to quantitate the metabolic fluxes for *E.coli* at different carbon sources [38]. Figure 1 shows the effect of dilution rate on the metabolic flux distribution for the case where acetate was used as a carbon source in the continuous culture. The metabolic flux analysis indicated that a slight change in flux distribution was observed for acetate metabolism even when the specific growth rate (dilution rate) was changed, indicating the subtle regulation mechanism exists in certain key junctions such as TCA cycle pathway and glyoxylate pathway. It was also found that different from acetate metabolism, pentose phosphate flux was significantly changed with the change in the specific growth rate when glucose was used as a carbon source due to the NADPH requirement (see Fig.2) [38].

Metabolic Flux Analysis with Gene and Protein Expressions

We recently used both ^{1}H-^{13}C 2D NMR and GC-MS to quantify the flux distributions of cyanobacteria at different culture conditions, and obtained some insight into the metabolic regulation with respect to culture environment [42]. The main metabolic pathways of cyanobacteria is shown in Fig.3. The fractionally ^{13}C-labeled *Synechocystis* sp. PCC6803 was harvested from the culture, subjected to hydrolysis, and the labeling patterns of the amino acids in the hydrolysate were analyzed using two-dimensional NMR spectroscopy and GC-MS.

The result shows that the flux of CO_2 reduction through the Calvin cycle was 211.4% of the glucose input flux. The reaction mediated by the fructose-1,6-bisphosphatase was found to be present in *Synechocystis* grown mixotrophically, demonstrating the presence of an ATP-dissipating futile cycle via ATP-consuming phosphofructokinase and fructose-1,6-bisphosphatase. The CO_2 fixation through the phosphoenolpyruvate carboxylase was 73.4% , which represented about 25% of the assimilated CO_2. The reaction catalyzed by the malic enzyme was identified by the labeling experiments, and the backward flux from the TCA cycle to glycolysis was found to be 84.6%. This explains the

242

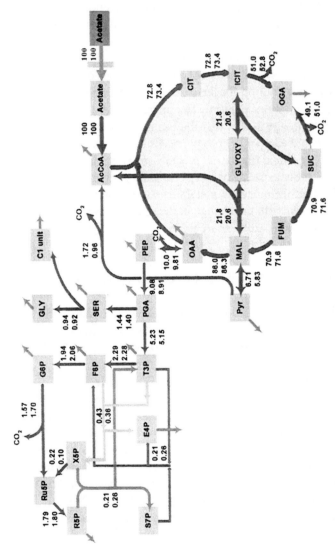

Fig.1. Metabolic flux distribution in acetate metabolism of *Escherichia coli* K12 in chemostat cultures at dilution rate of 0.11 (lower)and 0.22 h⁻¹, flux values at D of 0.22 h⁻¹ are at upper values and those for 0.11 h⁻¹ are at lower values.

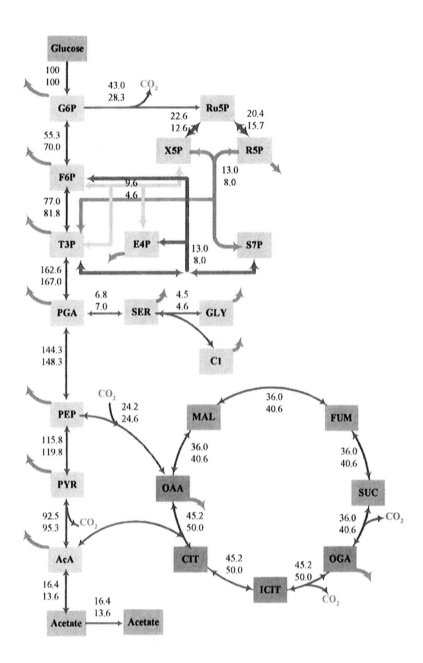

Fig.2 Net flux distribution in glucose metabolism of *Escherichia coli* K12 in chemostat cultures at D of 0.11 and 0.22 h^{-1}, flux values at D of 0.22 h^{-1} are at upper values and those for 0.11 h^{-1} are at lower values.

244

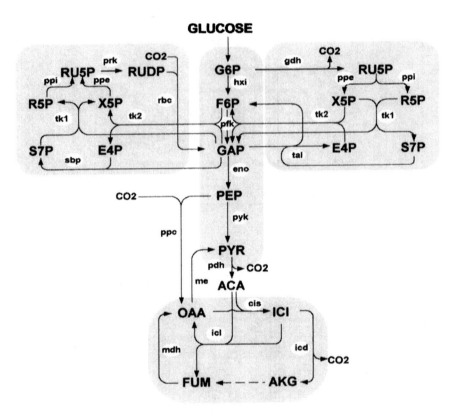

Fig. 3 Metabolic pathways of cyanobacteria.

significant increase of C2-C3 carbon bond cleavage in pyruvate when compared to the conservation of C2-C3 connectivity in PEP. High exchange rates in the glucose-6-phosphate isomerase, ribose-5-phosphate isomerase, glyceraldehyde-3-phosphate to phosphoenolpyruvate (PEP) conversions were found, and the PEP synthase were identified to be inactive during growth mixotrophically.

It should be noted that it is important to understand the regulation mechanism of gene and protein expressions as well as metabolic regulation. We, therefore, investigated how culture environments affect those regulations for *Synechocystis* using RT-PCR and 2-dimensional electrophoresis (2DE) as well as NMR and GC-MS [43]. We found that many genes are differentially regulated according to different mechanism, and the regulation of metabolic fluxes may be exerted at the transcriptional, post-transcriptional, translational, post-translational, and metabolic levels. Although at present, the transcriptomics, proteomics, and metabolic flux analysis allow high-throughput analysis of gene expression profiles, each of these techniques has its own advantages and limitation, and only their integration may provide us with a detailed gene expression phenotype at each level and allow us to tackle the great complexity underlying biological processes.

Table 1 shows the comparison on how gene and protein expressions as well as metabolic fluxes were changed between heterltrophic and mixotrophic conditions [43].

The enzyme of CO_2 fixation in Calvin cycle, ribulose bisphosphate carboxylase/ oxygenase (RubisCO), is encoded by the large (*rbcL*) and small (*rbcS*) subunit genes, which form an operon and are part of a single transcriptional unit. The result of gene expression shows that the transcript level of *rbcLS* decreased in the heterotrophic culture. From the 2DE result, the abundance of the S subunits (RbcS) was determined. As shown in Table 1, the relative protein abundance was consistent with the transcript levels of the gene *rbcLS* in the different cultures. Hence, the expression of gene *rbcLS* was transcriptionally regulated by light so that the synthesis of the enzyme was decreased about two-fold in darkness. However, when compared to the change in fluxes, it was found that only the transcriptional regulation was not enough and further regulatory mechanism is required to inactivate the synthesized enzyme in the heterotrophic culture (see Table 1). Therefore, besides the transcriptional regulation, the activity of RubisCO was also subjected to post-

Table 1 Changes in transcript levels, protein abundance and metabolic fluxes of *Synechocystis* in the heterotrophic (H) cultures compared to those in the mixotrophic (M) culture

gene	transcript ratio H vs M	protein ratio H vs M	flux ratio H vs M
rbcLS	0.5	0.5	0
prk	0.9	0.8	0
zwf	1.1	—	∞
devB	—	1.0	∞
gnd	1.6	—	∞
gap2	0.5	0.5	1.00
gap1	0.9	—	1.14
fbp	0.9	—	0
pfkA	1.0	—	12.7
fbaA	0.8	0.8	12.7
fda	—	0.9	12.7
glk	1.1		2.24
cfxE	—	0.8	2.00
ppc	1.0	—	1.70
icd	1.0	—	1.56
citH	1.1	—	1.78

translational regulation. Both regulatory mechanisms can cause an immediate cessation of CO_2 fixation during light-dark transitions.

In the three different cultures, the mRNA levels of the gene *prk* were similar, and the protein abundance was also roughly unchanged. In contrast, a significant flux difference was observed with the absence of light energy (see Table1). This indicates that the regulation of the flux through the phosphoribulokinase (PRK) was exerted after the enzyme synthesis.

The 2DE result shows that the synthesis of thioredoxin (TrxA) decreased about two-fold. The synthesis of ferredoxin (PetF) was not affected significantly, while the ferredoxin-thioredoxin reductase (FtrV) was down-regulated about five-fold in the darkness. Therefore, the absence of light energy probably resulted in the cessation of the ferredoxin/thioredoxin system, whereas this system was triggered by the light to recognize and reduce the specific disulfide bond of PRK for enzyme activation.

The transcript levels of the gene *zwf* were roughly unchanged under the different trophic conditions. Although the abundance of glucose-6-phosphate dehydrogenase (G6PDH) was not determined from the 2DE, the protein spot for the enzyme catalyzing the neighboring reaction, DevB, was identified. The synthesis of DevB was not affected significantly for the three cultures. However, the flux through G6PDH and DevB was activated in the heterotrophic culture (see Table 1). Thus it implies that the regulation of the flux through G6PDH and DevB was accomplished after the synthesis of the enzymes.

Synechocystis harbors two *gap* genes encoding distinct enzymes of glyceraldehyde-3-phosphate dehydrogenase, which appear to be essential for catabolic glucose degradation (*gap1*) or to operate in both the Calvin cycle and gluconeogenesis (*gap2*). From the amounts of total RNA required in RT-PCR experiments, it can be seen that the mRNA of *gap2* was more abundant than *gap1* transcript. We observed that the mRNA and protein expression levels for *gap2* were decreased about two-fold in the heterotrophic culture.

The fructose-1,6-bisphosphatase (FBPase) encoded by the gene *fbp*, is an important enzyme involved in gluconeogenesis and Calvin cycle. The FBPase in cyanobacteria also showed the sedoheptulose-1,7-bisphosphatase (SBPase) activity. From Table 1, it can be seen that the transcript level of the gene *fbp* was almost unchanged, while the flux was decreased to zero in the heterotrophic culture. For the reverse reaction catalyzed by phosphofructokinase (PFK), a significant increase in the flux was observed in the heterotrophic culture, but the transcript abundance was roughly unchanged (see Table1). Therefore, the results suggest that the activities of both FBPase and PFK were regulated at the metabolic level by the concentrations of the metabolites including substrates, activators and inhibitors.

The mRNA and protein levels for the genes *fbaA* and *fda* were similar under the different trophic conditions, but the flux through the fructose-1,6-bisphosphate aldolase (FbaA) increased significantly in the heterotrophic culture. This suggests that the regulation of the FbaA flux is not accomplished at the level of enzyme synthesis but affected via concentrations of substrates and products.

For the other genes involved in carbon metabolism including *glk*, *cfxE*, *ppc*, *icd* and *citH*, the transcript levels were almost unchanged for all cultures, and the fluxes were increased by 50-120%. This result is not surprising considering the variation in the intracellular levels of substrates and cofactors.

The above analysis is also under investigation for various *E.coli* cells in our laboratory.

Conclusions and Discussion

From the above analysis, it was found that the modulation of the fluxes in *Synechocystis* can be exerted through alterations in gene expression or through metabolic regulations. Therefore, the combination of the complementary information obtained from the mRNA expression, the protein expression and the metabolic fluxes is required to develop a better understanding of the regulatory events involved in the different trophic cultures of *Synechocystis*. The regulation studies based on the expression information at only one level (e.g., mRNA expression level) may lead to incorrect conclusions. For example, the post-translational modification of the cyanobacterial RubisCO during the light-dark transitions would not be identified based solely on mRNA and protein expression data. That is, the expression information at the mRNA, protein and metabolic levels must be collected together to develop a deep understanding of the interactions in a complex network.

This kind of research is important to grasp the cellular regulatory mechanism by taking into account three different levels such as mRNA and protein expressions and metabolic flux distribution in response to the culture environment. Here the semi-quantitative RT-PCR, 2DE and metabolic flux analysis by isotope labeling were used to determine the transcript levels, protein abundance and metabolic fluxes, respectively. Although the RT-PCR can be applied to detect the mRNA levels, the systematic analysis of expression phenotype requires more powerful technologies such as DNA microarray that enable high-throughput analysis of the gene expression at the transcriptional level. For the proteome analysis, further studies directed to the identification of the individual proteins will help to develop a better understanding of the global gene expression at the protein level. Recently, we measured 28 enzyme activities for the main metabolic pathway of $E. coli$, and compared with 2DE result. It was shown that the protein abundance obtained by 2DE was well correlated with enzyme activities except only a few proteins. Additionally, in order to analyze the intracellular fluxes in a more detailed metabolic network, it is necessary to measure as many labeled metabolites as possible and perform a thorough statistical analysis. More importantly, with the availability of massive amounts of expression information at the transcriptional, translational and metabolic levels, mathematical methods are required for integrating all the available information more efficiently.

Finally, it should be mentioned about the cell modeling in relation to the investigation on the metabolic regulation. Palsson and his co-workers [1,44] constructed an *in silico* strain of *E. coli* K-12 from annotated sequence data and from biochemical information. Using this *in silico* microorganism, one can study the relation between *E. coli* metabolic genotype and phenotype in the *in silico* knockout study [44]. The heart of this perspective is the study of the system as a whole rather than the detailed study of individual components and their direct interactions. Tomita et al.[45] developed an E-Cell model for the *in silico* experiment. We are now collaborating at Keio University for the development of quantitative modeling of *E. coli*. In particular, we are now integrating all the information obtained from gene expression, protein expression, and metabolic flux distribution as well as intracellular metabolite concentrations for the specific gene knockout *E. coli* cells.

Acknowledgement

It is acknowledged that this research was partly supported by the grant from New Energy and Industrial Technology Development Organization

(NEDO) of the Ministry of Economy, Trade and Industry of Japan (Development of a Technological Infrastructure for Industrial Bioprocesses Project).

References

[11]Schilling, C.H., J.S. Edwards and B.O. Palsson, Toward Metaboolic Phenomics : Analysis of Genomic Data Using Flux Balances, Biotechnol. Prog., 15, 288-295 (1999).
[2]Kao, C.M., Functional Genomic Technologies: Creating New Paradiums for Fundamental and Applied Biology, Biotechnol. Prog., 15, 304-311 (1999).
[3]Velculescu, V.E., L. Zhang, W. Zhou, J. Vogelstein, M.A. Basrai, D.E.J. Bassett, P. Hieter, B. Vogelstein and K.W. Kinzler, Characterization of the yeast transcriptome, Cell, 80, 243-251 (1997).
[4]DeRisi, J.L., V.R. Iyer and P.O. Brown, Exploring the Metabolic and Genetic Control of Gene Expression on a Genomic Scale, Science, 278, 680-686 (1997).
[5]Hatzimanikatis, V., L.H. Choe and K.H. Lee, Proteomics: Theoretical and Experimental Considerations, Biotechnol. Prog., 15, 312-318 (1999).
[6]Stephanopoulos, G., Metabolic Engineering," Curr. Opinion in Biotechnol., 5, 196-200 (1994).
[7]Bailey, J.E., Toward a science of metabolic engineering, Science, 252, 1668-1675 (1991).
[8]Stephanopoulos, G. and J.J. Vallino, Network rigidity and metabolic engineering in metabolite overproduction, Science, 252, 1675-1681 (1991).
[9]Cameron, D.C. and I.-T. Tong, Cellular and Metabolic Engineering An Overview, Appl. Biotechnol., 38, 105-140(1993).
[10]Nielsen, J.,Metabolic Engineering: Techniques for Analysis of Targets for Genetic Manipulations, Biotechnol.Bioeng.,58, 125-132(1998).
[11]Liao, J.C. and J. Delgado, Advances in Metabolic Control Analysis, Biotechnol. Prog., 9, 221-233 (1993).
[12]Shimizu, K., An overview on metabolic systems engineering

approach and its perspectives for efficient microbial fermentation, J.Chin. Inst. Chem.Eng.,**31**, 429-442(2000a).

[13]Shimizu,K, Metabolic pathway engineering: Systems analysis methods and their applications, in Adv.in Appl.Biotechnol. (ed. By J.J.Zhong),ECUST press, China (2000b).

[14]Stephanopoulos,G., A.A.Aristidou, and J.Nielsen, Metabolic Engineering: Principles and Methodorogies, San Diego,CA, Academic Press(1999).

[15]Lee,S.Y. and T.Papoutsakis (eds.), Metabolic Engineering, New York, NY, Marcel Dekker (1999).

[16]Vallino, J.J. and G. Stephanopoulos, Metabolic Flux Distributions in *Corynebacterium glutamicum* During Growth and Lysine Overproduction, Biotechnol. Bioeng., **41**, 633-646 (1993).

[17]Zupke, C., Stephanopoulos, G., Intracellular flux analysis in hybridomas using mass balances and in vitro ^{13}C NMR, Biotech. Bioeng., **45**, 292-303(1995).

[18]Marx, A., A.A. de Graaf, W. Wiechert, L. Eggeling and H. Shohm, Determination of the Fluxes in the Central Metabolism of *Corynebacterium Glutamicum* by Nuclear Magnetic Resonance Spectroscopy Combined with Metabolite Balancing, Biotechnol. Bioeng., **49**, 111-129 (1996).

[19]Mavrovouniotis, M.L., Computer-Aided Design of Biochemical Pathways, PhD Thesis, MIT, Cambridge, MA (1989).

[20]Pissarra, P.N. and C.M. Henriksen, Fluxmap. A Visual Environment for Metabolic Flux Analysis of Biochemical Pathways, Preprint of the 7th Int. Conf. On Comp. Appl. In Biotechnol., Osaka, Japan, 339-344 (1998).

[21]O'Leary,M.H., Heavy-atom isotope effects on enzyme-catalyzed reactions, In H.L.Schmidt, H.Forstel, and K.Heinzingler (eds.) Analytical chemistry symposia series, vol.11, Elsevier, Amsterdam (1982).

[22]Winkler,F.J., H.Kexel, C. Kranz, H.-L. Schmidt, Parameters affecting the $^{13}CO_2/^{12}CO_2$ discrimination of the ribulose-1,5-bisphosphate carboxylase reaction, pp.83-89, In H.-L. Schmidt, H.Forstel, and K.Heinzinger (eds), Analytical chemistry symposia series, vol.11, Elsevier, Amsterdam (1982).

[23]Wiechert W. and A.A. de Graaf, Metabolic Networks: I. Modeling and Simulation of carbon Isotope Labelling Experiments, Biotechnol. Bioeng., **55**, 101-117 (1997).

[24]Wiechert W. C. Siefke, A.A. de Graaf and A. Marx, Metabolic Networks: II. Flux Estimation and Statistical Analysis, Biotechnol. Bioeng., **55**, 118-135 (1997).

[25]Zupke, C. and G. Stephanopoulos, Modeling of Isotope Distributions and Intracellular Fluxes in Metabolic Networks Using Atom Mapping Matrices, Biotechnol. Prog., **10**, 489-498(1994).

[26]Schmidt, K., Carlsen, M., Nielsen, J., and Villadsen, J., Modeling isotopomer distributions in biochemical networks using isotopomer mapping matrices, Biotechnol. Bioeng., **55**, 831-840 (1997).

[27]Christensen,B., J.Nielsen, Isotopomer analysis using GC-MS, Metabolic Eng.,**1**, 282-290(1999).

[28] Schmidt, K., Nielsen, J., and Villadsen, J., Quantitative analysis of metabolic fluxes in *Escherichia coli*, using two-dimensional NMR spectroscopy and complete isotopomer models, J. Biotechnol., **71**, 175-190 (1999).

[29]Mollney, M., W. Wiechert, D. Kownatzki and A.A. de Graaf, Bidirectional Reaction Steps in Metabolic Networks: IV. Optimal Design of Isotopomer Labeling Experiments, Biotechnol. Bioeng., **60**, 86-103 (1999).

[30]Wiechert W., M. Mollney, N. Isermann, M. Wurzel and A.A. de Graaf, Bidirectional Reaction Steps in Metabolic Networks: III. Explicit Solution and Analysis of Isotopomer Labeling Systems, Biotechnol. Bioeng., **66**, 69-85 (1999).

[31]Wiechert W., M. Mollney, S.Petersen, and A.A. de Graaf, A universal framework for ^{13}C metabolic flux analysis, Metabolic Eng., 3, 265-283(2001)

[32]Szyperski ,T., Biosynthetically directed fractional ^{13}C-labeling of proteinogenic amino acids: An efficient analytical tool to investigate intermediary metabolism, Eur.J.Biochem., **232**, 433-448 (1995).

[33]Arauzu, M. and K.Shimizu, An improved method for the metabolic flux calculation based on isotopomer distribution with analytical expressions, J.Biotechnol., submitted (2002).

[34]Lee,W.-N. P., L.O. Byerley and E.A.Bergner, J.Edmond, Mass isotopomer analysis: Theoretical and practcal considerations, Biol. Mass Spectrometry, **20**,451-458 (1991).

[35]Lee,W.-N. P., E.A.Bergner, Z.K.Guo, Mass isotopomer patterns and precurser-product relationship, Biol. Mass Spectrometry, **21**,114-122 (1992).

[36]Wittmann, C., and E.Heinzle, Mass spectrometry for metabolic flux analysis, Biotech. Bioeng., **62**, 739-750 (1999).

[37] Yang,C., Q.Hua, K.Shimizu, Quantitative analysis of intracellular metabolic fluxes using GC-MS and two-dimensional NMR spectroscopy, J.Biosci. Bioeng., **92**,277-284(2002).

[38] Zhao, J. and K.Shimizu, Metabolic flux analysis of E.coli K12 grown on 13C-labeled acetate and glucose using GC-MS and powerful flux calculation method, J.Biotechnol., in press (2002).

[39]Klapa, M.I., S.M. Park, A.J. Sinskey and G. Stephanopoulos, Metabolite and Isotopomer Balancing in the Analysis of Metabolic Cycles: I. Theory, Biotechnol. Bioeng., **62**, 375-391 (1999).

[40]Park, S.M., M.I. Klapa, A.J. Sinskey and G. Stephanopoulos, Metabolite and isotopomer balancing in the analysis of metabolic cycles, Biotechnol.
Bioeng., **62**, 392-401 (1999).

[41]Christensen,B., J.Nielsen, Metabolic network analysis of *P. chrisogenum* using ^{13}C-labeled glucose, Biotechnol. Bioeng., **68**, 652-659 (2000).

[42]Yang,C., Q.Hua, K.Shimizu, Metabolic flux analysis in *Synechocystis* using isotope distribution from ^{13}C-labeled glucose, Metabolic Eng. , in press (2002).

[43]Yang,C., Q.Hua, K.Shimizu, Integration of the information from gene expression and metabolic fluxes for the analysis of the regulatory mechanisms in *Synechocystis*. Appl. Microbiol. Bioeng., **58**,813-822 (2002).

[44] Peng,L. and K. Shimizu, Global metabolic regulation analysis for *E.coli* K12 based on protein expression by 2DE and enzyme activity measurement, Appl.Microbiol.Biotech., in press (2002).

[45]Edwards, J.S. and B.O. Palsson, *Escherichia coli* K-12 *in silico*: Definition of its metabolic genotype and analysis of its capabilities, Submitted for publication (2000).

[46] Tomita,M. et al., E-Cell project (2002).

[47]www.ttck.keio.ac.jp/IAB

Process Validation

Chapter 16

Fermentation Process Validation

D. Cossar

Cangene Corporation, 3403 American Drive, Mississauga,
Ontario L4V 1T4, Canada

A discussion of process validation relating to fermentation is presented. The importance of baŝis in understanding of the production organism physiology in response to physicochemical process control parameters is emphasized. Validation is relevant to all stages of process development and is an ongoing activity requiring company wide input.

Introduction

Validation is an integral component of current Good Manufacturing Practices (cGMP), as defined in the FDA Code of Federal Regulations (CFR 21 Parts 210 and 211). Industry guidelines are available in ICH document Q7A – Good Manufacturing Practice Guidance for Active Pharmaceutical Ingredients (2001). Validation applies to all the elements involved in production of a therapeutic substance (otherwise known as Active Pharmaceutical Ingredient or API). This includes facility maintenance and operation; equipment maintenance and operation; cleaning; sterilization; analytical methods; and the process itself. Most of these systems have been well established in the chemical drugs industry and readily transfer to the biotechnology field. There are many sources of guidance on validation – from the FDA or ICH documents referred to above to published guides (*1*). There are only a few which discuss how process validation is actually accomplished (2). Most of the available documents indicate that validation of processes must be carried out but provide little pragmatic information as to how this might be achieved. The main problems arise when process validation is considered for fermentation-based manufacturing.

It is well understood that biological production systems, as applied for recombinant DNA derived products, are perceived to be inherently variable. The microorganisms which form the basis of the process will adapt in a transitional manner to changes in the physicochemical environment. It may equally be argued, from a physiological viewpoint, that biological processes are actually inherently stable. They will tend to do the same things with a substrate over a comparatively broad range of conditions and genetic adaptation through mutation can take a large number of generations to establish. However, the burden of "proof" remains to demonstration of this stability. Products derived from biologically-based processes are regulated under the auspices of the FDA's Center for Biologics Evaluation and Research (CBER). The current philosophy is that this class of drugs cannot rely solely on final product testing for confidence of patient safety. This thinking is rooted in the issues of biological complexity and the limited sensitivity of analytical techniques. The end result is that drugs derived from biological production processes rely on demonstration of process consistency for safety consideration (the "product by process" doctrine). Process validation thus becomes the primary tool applied to regulate the biotechnology industry.

Validation can be prospective – where it is conducted before a process is used in manufacturing; concurrent – where it is conducted alongside manufacturing; or retrospective – where the manufacturing process is in existence and is ongoing. The ideal is to conduct prospective validation followed by concurrent process monitoring rather than to attempt to establish a manufacturing process first.

Process Validation

Process validation is defined as:

> "*Establishing documented evidence which provides a high degree of assurance that a specific process will consistently produce a product meeting its pre-determined specifications and quality attributes.*"[1]

---

[1] USFDA Guideline on General Principles of Process Validation, May 1987, p4.

The key words in this definition are "documented" and "consistently". It is essential that validation be properly documented and conducted under appropriate levels of control. Consistency within a specified matrix of operational ranges according to a controlled set of manufacturing instructions is the ultimate goal for a validated process. These operational ranges or limits must be demonstrated to encompass parameter values which do not have significant impact on product safety. In terms of fermentation, at least, it is the requirement for "demonstration" which is the most problematic. There is almost an inconsistency between the concept of supporting validation outcomes by analysis of end product and the "product by process" approach to drug approval. However, there is a limit to how much clinical testing (the ultimate "proof" of product safety) that can be done and validation will provide information as to the response surface where risk is minimized. Validation should not be considered as a discrete event, conducted once and subsequently archived. GMP requires ongoing monitoring of processes as an essential component of compliance. This monitoring should be applied using classical trend analysis methods – especially to permit identification of a potential problem before process failure and product rejection. Such monitoring also requires periodic scientific review of data, ideally by members of the Research and Development group responsible for development of the process. Validation thus becomes a company-wide activity.

In classical terms, validation involves manipulation of one variable at a time (be it the preparation of a buffer for an assay, the load to an autoclave, the concentration of a disinfectant or cleaning agent, for example), demonstrating that the outcome of the procedure is the same, and showing that there is no adverse impact on the product. This approach is manageable where the number of parameters to be evaluated is small and the interaction between variables inconsequential. In terms of fermentation, such a structured approach rapidly runs into the problem of number of experiments required. The product of a fermentation step in a process is often a relatively crude mixture of components which is difficult to analyze in a meaningful, quantitative, manner. In addition, if final product quality becomes the yardstick, the human resources involvement is unsupportable for any but the largest and most profitable manufacturers. This is both because the number of variables is large and because they are interdependent. It is therefore necessary to reduce the number of variables to be tested to a minimum whilst retaining an understanding of the manufacturing plant reality. It is evidently possible to set all operational limits for a process as being the same as the practical limits of control imposed by the process equipment. Such an approach has high safety profile but will also result in a high number of rejected processes. There is thus a trade-off between definition of process ranges which reduce risk of failure and their demonstration as acceptable ("safe").

The preferential entry point to validation for fermentation is the development of the process (3). During this stage, the important parameters are defined and the preferred values determined. The capacity of the purification process to purify product away from host-derived components is normally determined concurrently with fermentation. Finally, the analytical tools to characterize the product through the process are developed in response to experience with the vagaries of the product itself in context of the production system. The prime importance of good documentation during process development is thus highlighted. The end result is a body of information surrounding a bench-scale process which provides product of a certain quality. The bench-scale process is then taken to pilot or full manufacturing scale prior to testing in animal models for toxicity and subsequently into Phase 1 clinical trial for human safety. This provides the baseline process profile with defined transfer points between stages and a series of in-process control analyses which is the benchmark for validation.

The question of in-process control testing is also an important consideration, since these are included in the quality indicators used in demonstrating successful validation. The process development phase (including scale-up) should be used to gather data to allow elimination of variables from the list of those items requiring validation. During development, there is a tendency to maximize the number of tests in a process – partly on the assumption that this will improve control. These tend to be carried through to scale-up – mainly to provide confidence of successful transfer from bench to pilot or production scale. However, each of these must be considered during validation. Thus the preferred approach is to monitor a large number of variables during process development and to identify those which are critical indicators of process control. The majority of the tests can thus be dropped from validation to monitoring status. In this regard, a good method need only be precise. Variability in an analytical method can only compound the difficulties of validating a potentially variable process.

Once the critical process parameters have been identified through the development programmed a validation protocol is written. The goal of this is to establish a series of experiments to test the limits for these parameters – individually and in combination. There is an important caveat at this point – numerology dictates that the number of combinations to be tested increases dramatically with the number of individual cases involved. A scientific case may be made to eliminate many of these combinations as insignificant, statistically unlikely, or physiologically implausible. For those parameters which remain, the validation will ideally, but not necessarily, demonstrate the failure limits of the process as well as the safe limits and the operational limits to form a nested set. That is where the operational (control) limits are encompassed by the range of values shown to have no evident impact on product (safe limits) and which, in turn, lie within those conditions which result in process failure. Thereby providing a high degree of confidence that the process will consistently provide a product which is safe for therapeutic use. The "good scientific argument" scenario may be coupled with a definition of "worst case" conditions to further reduce the number of experiments required. Thus, for example, the upper limit of parameter "A" may have more potential consequence on product quality than the lower limit. It may therefore be possible to define a lower safe limit and both safe and failure limits for the high ends of the range. For other parameters, there may be no discrete "failure" limits – for example, process temperature, within a relatively broad range, may be considered to simply reduce the rate of the biochemical processes occurring within the cell. Thus the process will arrive at a certain point at a time which is dependent on the temperature. Assuming that this point is the same (i.e. a target biomass for culture harvest) then the process can be considered not to have failed over this entire range. However, in the real world, process time in a manufacturing plant is costly and so the process will be operated under conditions designed to get the purified product in the shortest reasonable time. Thus the "failure" limits can be considered to be those outside of which the process takes too long or is too short to fit into the plant schedule. There is no rational reason to define or validate these ranges. However, experience in a manufacturing environment quickly shows that process controllers do malfunction and this can result in "drift" of a control variable and it is essential, from a pragmatic perspective, to have evidence which shows that this deviation has no adverse effect on the product.

Arguably one of the most important sections of the validation protocol is that which defines the success or failure of the process. The product of a fermentation process is, at best, a process intermediate where the ultimate product exists in a soup of "background" contaminants. It is often difficult, if not impossible with current methods, to evaluate the quality of the product at this stage of the overall manufacturing process. The quantity, or yield of product may be an acceptable surrogate for most experiments. It remains impossible to predict the effects of changes in fermentation on the profile of contaminants through the recovery and purification stages to the final drug substance. That is without actually doing the purification. It may not be necessary to carry all validation experimental processes through to final drug substance since some of them may fail at an early stage and it may be possible to use normal in-process controls and specifications as acceptance criteria. This latter case is, to a certain extent, based on overcapacity built in to the process. Thus, for example, a chromatography step will not generally be conducted at maximum resin capacity – there will be some binding capacity of resin left unused with the normal process stream. Changes in the load to the column, which may arise through changes in the upstream fermentation, are accommodated by this excess capacity and the process runs normally thereafter. However, it is good practice to retain archived samples to permit evaluation of at least the preliminary stages of purification for all validation experiments. It is obvious that the majority of this validation should be conducted at bench-scale insofar as this can be claimed to be representative of the full scale process. For some manufacturing systems, it may be required to conduct some validation at pilot or manufacturing scale. In any case, validation cannot be completed until the process has been conducted successfully in the manufacturing plant on a minimum of three consecutive occasions. It is equally important in this context to consider conditions under which re-validation is required. Thus changes in manufacturing facility (including equipment, environment, procedures, and materials) must be evaluated in terms of potential impact on the process and, if necessary, tested before implementation.

Fermentation Process Validation

The detailed approach to fermentation process validation is dependent on the nature of the process – whether it is based on intra- or extracellular accumulation of product; whether the producing strain has particular sensitivities or characteristics; and the characteristics of the product itself. The validation plan must be prepared with an understanding of how the fermentation works. The questions of how and why the system responds in a particular manner should be considered and, ideally, understood. This absolutely requires consideration of the physiology of the production strain, since physiological state defines the chemical composition of the cell and therefore the profile of materials transferred to the purification process. From this perspective, the validation approach can be constructed and, more importantly, defended on regulatory review.

A representative validation scheme can be developed for a classical recombinant DNA protein product. This assumes a three stage fermentation from seed flasks through an intermediate fermentation step to a fed-batch production fermentation. The product accumulates inside the cell in response to a specific induction event. Transfer from stage to stage is determined by time and/or biomass-related criteria. The fermentation is operated with dissolved oxygen and pH control. Significant divergence of other common types of process from this model will be discussed where relevant. The issue of scale must also be considered – there are inescapable consequences for increase in volume of the bioreactor (4). These include mixing times, shear, and heating/cooling profiles – all of which can impact the culture physiology and fermentation process. The preference is to conduct as much of the validation at bench scale as possible to minimize cost. However, the approach must also include critical evaluation of process scale to demonstrate why this is appropriate in context of the particular process.

In any industrial fermentation process there are a number of "control" variables and "response" variables which are used to direct the process through to the harvest point. The two are not necessarily distinct since a response variable can be used for process control. However they both must be considered for validation purposes. The control variables are those which are independent of the process – raw materials, process timing, temperature, pH, air flow rate, impeller agitation speed, for example – those which the operator inputs to the system. The response variables would be biomass (whether this be optical density, dry weight, etc) or culture physiology (carbon dioxide production rate, oxygen consumption rate, respiratory quotient, substrate consumption, or metabolite production for example). It is comparatively trivial to manipulate the control variables but equally difficult to manipulate the response variables – especially in isolation.

In practical terms, the control variables form the place to start validation. Each of these has an effect on the fermentation process at some level and they are discussed in the following sections and in the context of the process stage.

Main Parameters for Validation

Raw Materials

The validation process should include evaluation of the influence of media formulation on the process. This need not extend to a complex array of experiments to establish limits for each component – primarily this should establish that the contributions of the components to growth are understood and that any growth limiting components are identified. Supportive data on media composition can be obtained by determining yield coefficients for the various components (principally the macro-nutrients) during process development. The majority of factors would be expected to be in excess for standard batch processes, but not to the point of being toxic or growth retardant. Hence variability arising from minor changes in concentration (from balance accuracy, for example) is not significant. Where the process involves defined media with well-characterized components, it is arguable that variability is unlikely to arise from the use of different manufacturers lot of raw materials. However this may not be the case where the media formulation includes complex, ill-defined, factors such as protein hydrolysates or crude extracts of plant origin (see figure 1 for example). In this scenario, it is important to evaluate effects of using different batches of the more complex items on the growth of the production organism as early as possible in development. In dealing with raw materials it is important to consider the way in which these are handled prior to introduction of the microorganism. Thus issues of shelf life (as powder and in solution), hold time prior to sterilization, the sterilization process itself, and even hold time pre-inoculation should be assessed for potential impact on process. It may also be important to consider the specifications for raw materials release. Typically, these are compendial (against USP or EP standards, for example) and based on chemical identity. There may be specific characteristics which should be introduced to allow for biological systems. These may include such tests as bioburden (very high levels may swamp capacity to sterilize media); endotoxin (which is of real concern for injectable drug products); or growth-affecting substances (with both positive or negative consequence). Sourcing of materials during the development phase should be done in cognizance of the manufacturing reality. When using small quantities of chemicals for bench-scale processes, it is possible to use high purity materials which are uneconomically expensive for production scale. Conversely, bench-scale processes may use water, steam, and air of a lower quality than typically encountered in a manufacturing environment. Such metabolic flux pathways. However, there is unlikely to be a radical shift in enzyme complement or in macromolecular composition of the cell. Effects on product quality may be considered to be unlikely – at least in terms of contaminant profile of purified product. This phenomenon may also influence the uptake of an inducer – affecting product expression rate and ultimate process yield. The direct influence of pH on microbial growth is to decrease the specific growth rate on either side of an optimum. The rate of decline is organism specific and should ideally be determined for the production organism and under conditions of temperature and medium formulation as close to those for production as possible. Most often, the rate of decline in growth rate is lowest around the optimum pH value for the organism. This is the preferred condition for validation – where the largest pH range can be issues may have process impact that is not evident until the process is transferred into manufacturing with consequent costly failure. The optimal situation is to use identical raw materials during development and manufacturing.

pH

Microorganisms react to a change in the pH of the external medium in a manner directed to maintenance of a constant intracellular pH. Transport of substances across the cell membrane is known to be pH sensitive, more specifically by pH gradient between the cytoplasm and the culture medium. This can influence the composition of the cell by altering the balance of tolerated before the process falls out of the biomass/time specification. The culture pH can also have dramatic effects on product where this is secreted from the producing cell (for example through interaction between product and protease). For such systems it may not be possible to conduct the process at optimal pH for growth and a tighter pH specification may result. It is evident, given the multiple possible effects, that pH is a parameter requiring validation.

There are two scenarios with pH - that it is controlled at a defined value by metered addition of acid and/or alkali or that it is not controlled other than by the buffering capacity of the medium. In general, and in absence of organism-specific factors, it is arguably better from the validation perspective to control pH at a defined value. Where pH is allowed to drift from the initial value the question for validation is one of ensuring constant profile within specified limits. There are therefore two variables involved – actual pH and time in the process. In this case, validation may consist of adjusting the initial

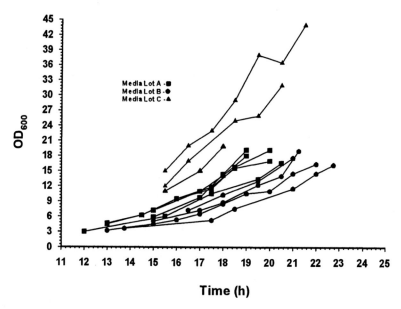

Figure 1. Growth profiles of a series of fermentations using media prepared using different lots of casein peptone from a single manufacturer. Streptomyces lividans cultures were grown in 15 L stirred tank bioreactors (Chemap AG, Switzerland) in glucose-Casein peptone-salts medium at 32 °C, pH 6.7, >25% Dissolved oxygen.

pH of the culture medium and then running the process. Analysis of potential risk is complicated by the fact that effects on the system are not constant – the organism is continually adapting to a new environment. There is also the relationship between pH and buffering capacity – being inversely related to difference between actual pH and pK_a of the system. Hence the pH profile may change significantly through a minor change in initial value. The situation where pH is controlled at a defined value is comparatively simple. The microorganisms experience a more consistent environment and validation at pH values on either side of the control value is a single parameter case. However, the nature of system controllers is that they sometimes misbehave and step changes in pH can occur. It is not necessarily appropriate to claim that a step change within a validated range is acceptable since the shock effect on cell physiology cannot be ignored. In this case, the suggested approach is to test a range of pH values which is linked to the buffering capacity of the medium. Where this is high, a relatively broad range can be applied on the argument that changes in pH within the range are likely to be slow and hence shock effects minimized. Conversely, if the buffering capacity is low, then a narrow range (but still outside the operating range of the pH control systems in the processing plant) should be applied. Evidently, media formulation development should move towards high buffering capacity around the preferred process pH value. Overall, there is a balance between a better controlled (and validatable) system which is prone to higher failure rate to one which offers lower control confidence but also lower risk of failure.

Temperature

As for pH, temperature has a primary effect on cell growth rate and most probable cause of process failure is concerned with the biomass/time control parameters. The pattern of response is slightly different from pH in that fall off in growth rate above the optimum is frequently found to be more dramatic than at temperatures below it. Sub-optimal temperatures may be argued to show lower growth rate through thermodynamic effects on rates of reaction. Super-optimal temperatures on the other hand may cause a lowering of growth rate through protein denaturation effects – with consequences for cell composition (primarily a higher content of missfolded protein). It may therefore be argued to be more important to establish the safe and failure limits for temperatures above the process control value. It is recommended to acquire information on the growth rate/temperature relationship for the production strain as part of the development programmed.

Experiments have shown that cellular composition (protein, DNA, and RNA) for *Salmonella typhimurium* did not change with temperature of incubation, despite a significant reduction in growth rate (*5*). Herbert (*6*) showed that this is true only for the case in batch culture where the organism is growing at the maximum possible growth rate under the imposed conditions. There must therefore be some caution applied to arguing lack of significant temperature effect on cell composition. In a fed-batch process where growth rate is controlled at a value lower than the maximum by rate of addition of a specific nutrient, there may be a change in RNA content with temperature at fixed feed rate (and vice versa – with flow rate at fixed temperature). These effects may be small but should not be ignored when planning validation experiments. Of some concern would be a process where feeding is defined to commence at a particular time post-inoculation. When temperature of the process is manipulated, the feed initiation will happen at a different biomass and this may affect the culture growth (by over- or under-feeding respectively). Secondly, the feed rate will impose a particular growth rate (in relation to the maximum possible at that temperature, assuming that one component is limiting) and this can alter the physiological status of the organism.

Temperature (and, indeed, any parameter which affects growth rate) must also be considered in context of mass transfer phenomena in the bioreactor. As temperature and growth rate increase, metabolic rate increases. The demand for nutrients (and to expel waste) will increase and the capacity of the system to effect transfer from the bulk liquid to and from the cell surface may become growth-limiting. The cells may become energy-starved (or poisoned by accumulation of metabolic by-products) and can significantly shift their composition (and hence potential contaminants to the purified product). The outcome can be process failure at sub-optimal temperature. There is a compounding effect specific to temperature since oxygen solubility decreases with increasing temperature. The capacity of the system to deliver oxygen to cells thus drops off significantly with increasing temperature.

There may be a particular issue with temperature for some commonly used approaches to control of gene product expression. In processes which use the γ_{pl} system, temperature is used to control activity of a promoter. In this case, there is a protein (repressor) which binds to the promoter and prevents interaction with RNA polymerase. As temperature is increased, the conformation of the repressor changes to a non-binding competent state, allowing access of RNA polymerase and transcription of the gene product. The issue of concern is that the transition from binding competent to binding incompetent is not a precise value – there is a range of temperatures over which this occurs which is affected by cell composition (primarily pH and ion concentration). The consequences of increasing temperature of incubation may thus be to "leakiness" at the promoter and premature product expression. This can have process consequences through product toxicity, energy or anabolic component diversion, or attenuation of promoter activity during the specific induction phase. In this system, the effects of scale on temperature control must also be considered – for example it may take longer to ramp temperature to achieve the binding incompetent state at manufacturing scale. The potential for this should be built in to small scale development by evaluating effects of rate of increase of temperature (or ramp slope) on the product yield and quality.

Hence a simplistic approach to validation of process temperature must be tempered by consideration of other components of the process.

Air Flow Rate and Impeller Agitation Speed.

These two parameters are considered together since they are inter-related, particularly in aerobic processes. The geometry and operation of the impellers in a stirred tank bioreactor have a major impact on mixing capacity of the system. They also define, to a large extent, the shear forces acting on the microorganisms in the process. Bulk mixing of a fluid is a product of the energy input to the system – more energy providing a shorter time to homogeneity following a perturbation. The energy input to a stirred tank bioreactor is provided via the impellers which act to accelerate the fluid. The baffles create turbulence in the fluid to enhance mixing. These accelerative and turbulent flow effects produce shear forces acting on the components of the system – including the microorganisms. In conjunction with air flow rate (and head space pressure) impeller setup controls gas transfer characteristics of the bioreactor. Microorganisms vary widely in their sensitivity to shear and their demand for gas exchange, and sometimes even over the course of a process. Thus unicellular bacteria are probably the least shear-sensitive and eukaryotic cells the most, with filamentous organisms somewhere in between. It is important to investigate the effects of shear on the process – especially since cell damage and release of intracellular components can have dramatic effects on an extracellular product (both by action of proteases and by increased contaminant concentration) as well as affecting growth rate. Manipulating shear in isolation is difficult because of the effects on oxygen transfer. At high shear there is inevitably better gas exchange between the air/liquid interface and at the bacterial surface. This is both advantageous and deleterious since oxygen is required for energy generation but is toxic at relatively high levels through production of superoxides and free radicals. This can impact both growth rate and cell physiology (production of protective proteins, for example) resulting in process failure. It is possible to manipulate the gas composition to decouple dissolved oxygen and impeller agitation speed but this requires specialist equipment. It is more convenient to consider effects of impeller agitation speed along with variations in dissolved oxygen. In this case, high range dissolved oxygen setpoint also covers high shear (and faster mixing times) and low range with low shear and poor mixing. In processes without oxygen control impeller agitation speed becomes the validation parameter with dissolved oxygen as a secondary (dependent) variable.

Biomass/Time

Biomass accumulation and time are frequently utilized process control parameters. This is encountered as an instruction to transfer culture from one stage of the process to another at a particular biomass or a particular time. This being on the premise that growth rate and culture physiology are effectively synonymous. The experiments reported by Herbert (6) show that this premise is reasonable where the culture growth rate is defined by rate of supply of a particular nutrient (i.e. in fed-batch or semi-continuous growth). It is also reasonable to assume that, all other parameters being consistent, the microorganism will grow at a particular rate and have a particular physiology. The important caveat may be the case of temperature or pH where growth rate may be lower or higher but the organism will have essentially the same physiological status since it is growing at the maximum rate possible under the circumstances and neither would be expected to impose a particular metabolic condition. However, this relates to interpretation of validation and not validation per se.

Both biomass and time would have ranges associated with them to allow for normal (acceptable) process variability (see figure 2 for example). The rationale is that a process is "normal" if the organism displays a particular growth profile and anything outside of this is aberrant and should be rejected. These ranges must therefore be validated. The problem is that growth rate is not a factor which can be tightly controlled by the operators. It is generally not possible to run a process to a specified value of either time or biomass independently of the other. Process development involves evaluation of optimum biomass to achieve a particular outcome for product recovery and purification. This is generally within a timeframe convenient to the operator's schedule. The issue with biomass determination is that of precision of the method used. Optical density is less precise because it often involves significant sample dilution to remain within the linear response range of the spectrophotometer, because time taken between sampling and analysis is variable and conditions are uncontrolled; and it is sample handling dependent (for example cells can settle, introducing non-homogeneity of sample). On-line, real time methods such as biomass-specific sensors or metabolite production (especially CO_2) are preferred, but may not be universally applied – for example in the shake flask stages of a process.

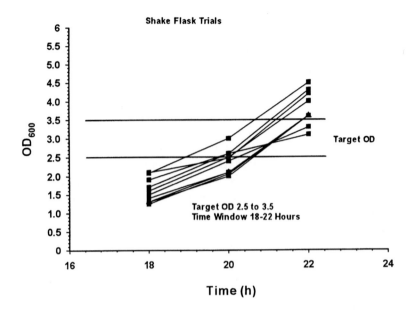

Figure 2. Growth profiles for 10 200 mL shake flask cultures of Streptomyces lividans grown in 1.0 L Erlenmeyer flasks with baffles in glucose-casein peptone-salts medium at 32 °C and 240 rpm in a New Brunswick Series 25 Incubator shaker.

Validation of the biomass/time process control parameters is probably only possible concurrently with process development. During this phase, a database of information can be obtained on what is the range of values for time to achieve a specified biomass when the process is conducted normally. Time, by itself, is sometimes used to define when, during a process, to do something – for example start a feed, or induce expression. The preferred option in this case is to substitute a biomass-related parameter for time. In the absence of an appropriate indicator other than time, it is essential to validate a range to demonstrate proper process control. This is in the context of normal growth profile variability and hence biomass at the defined time of process.

Conclusion

In conclusion, validation of a fermentation process involves consideration of many inter-related parameters. Some of these are operator-specified and some concern the response of the organism to the imposed conditions. The preferred basis for design of a validation protocol is a sound understanding of the physiology of the organism. This understanding is acquired during the development phase and can be applied with good scientific rationale to address the real potential for process failure.

References

1.   *A guide to Good Validation Practice.*; Kanarek A.D.; D&MD Publications, Weestborough, USA, 2001.

2.   Shillenn, J.K. Ed. "Validation Practices for Biotechnology Products." ASTM Publication STP 1260, USA, 1996.

3.   Cossar, J.D. Current Opinion in Drug Discovery and Development **2001**, 4(6), 756-759.

4.   Prokop, A. Adv. Appl. Microbiol. **1995**, 40, 155-236

5.   Schaechter, M.; Maaloe, O.; Kjeldgard, N.O. J. Gen. Microbiol. **1958,** 19, 592-606

6.   Herbert, D. Symp. Soc. Gen. Microbiol. **1961,** 11, 391-416

Indexes

Author Index

Subject Index

A

Acarbose, carbohydrate-based
therapeutic, 91t, 99

Acetate production and growth in cell
extracts production, *Escherichia
coli*, 148–152f

Acetate production disruption by gene
inactivation in *Clostridium
tyrobutyricum*, 57–58

Acetate production in butyric acid
formation, 54

Aeration rate, effect on fermentation
in rotating fibrous bed bioreactor,
47–49

Agitation rate effects, stirred tank
cultures using microcarriers, 135–
136f

Air flow rate, main process validation
parameter, 268–269

Alginate, production, overview, 11

Amino acids, production by
fermentation, 186–187f

5-Aminolevulinic acid and vitamin B_{12}
production in genetically engineered
Propionibacterium freudenreichii,
221–232

5-Aminolevulinic acid biosynthesis,
pathways, 222, 224f–225f

Amphotericin B, carbohydrate-based
therapeutic, 91t, 98, 103

Anaerobic fermentation, kinetic
model, 22–25

Anaerobic microorganisms,
continuous fermentation system,
21–35

Analytical methods
butyric acid production, 57
L(+)-lactic acid production, 41

Aspergillus candidus, filamentous
fungus, mannitol production, 76–77

Aureobasidium pullulans, yeast-like
fungus, mannitol production, 76

Azo dyes
bioprocess development for
bacterial decolorization, 163–170
industrial uses, 163

B

Bacterial decolorization, azo dyes,
163–170

Bacterial mercury resistance, 160–
162

Biofilm systems, fixed-bed bioreactor
with immobilized cells, bacterial
decolorization, azo dyes, 168–
169f

Biomass determination, method
precision in process validation,
269

Biomass from carbon dioxide
photosynthesis, sago palm, 32

Biomass/time, main process validation
parameter, 269–270

Bovine serum, contamination with
viruses, 97

2,3-Butanediol, production by
fermentation, 12t

Butyrate tolerance, *Clostridium
tyrobutyricum* mutants, 62–64

Butyrate tolerance study, 57

Butyric acid
applications, 53
production   enhancement by
Clostridium tyrobutyricum, 52–66
synthesis from petrochemicals, 53

Ethanol production
  catabolic pathways, *Rhizopus
    oryzae*, 37–38*f*
  fermentation, overview, 4–6
  Zymomonas mobilis, 31–32
Excretory production into culture
  medium, recombinant proteins,
  Escherichia coli, 202
Exopolysaccharides, production by
  fermentation, overview, 11
Expression vector development in
  propionibacteria, 223, 226–227*t*
Extracellular polysaccharides,
  biosynthesis by *Ganoderma
  lucidum*
  production, 114, 116, 118–119*f*
  proposed pathway, 109–110*f*
Extract preparation and protein
  synthesis in cell extracts production,
  Escherichia coli, 147–148

F

FDA Center for Biologics Evaluation
  and Research, 258
FDA Code of Federal Regulations,
  Good Manufacturing Practices, 257
Fed-batch fermentation, butyric acid
  production by *Clostridium
  tyrobutyricum*, 58–63
Fed-batch processes with high cell
  density fermentation, bacterial
  decolorization, azo dyes, 165
Fed-batch strategies in mercury
  detoxification operations, 162, 170
Fermentation
  carbohydrate-based therapeutics
    production, advantages, 95*t*
  cellular engineering for enhanced
    protein production, 193–206
  commodity chemicals production,
    overview, 3–17
  comparison between bacterial and
    fungal lactic acid methods, 37*t*

improvement by metabolic
    engineering, 235
  See also Continuous fermentation
    system for anaerobic
    microorganisms
Fermentation conditions in cell
  extracts production, *Escherichia
  coli*, 145–146
Fermentation kinetic study, butyrate-
  producing mutant development, 56
Fermentation kinetics
  butyric acid production by
    Clostridium tyrobutyricum, 59–63
  L(+)-lactic acid production by
    Rhizopus oryzae, 42–49
Fermentation process development,
  carbohydrate-based therapeutics,
  89–107
Fermentation process validation, 262–
  263
Fibrous bed bioreactor
  butyric acid production, 52–66
  construction and fermentation, 56
Filamentous fungal morphology
  control, immobilization, 36–51
Filamentous fungi, microbial
  production, mannitol, 76–77
Fixed-bed bioreactor containing
  immobilized cells, bacterial
  decolorization, azo dyes, 168–169*f*
Flux vector calculation, metabolic flux
  distribution, 235–236
Foldases for soluble protein
  production in periplasm,
  Escherichia coli, 201*t*
Folic acid in American diets, 208
Folic acid production
  chemical synthesis, 208
  metabolic engineering, 207–219
  stoichiometric model and methods,
    210–212
Food and Drug Administration. *See*
  FDA
Fraxinus ornus bark, mannitol source,
  68

Lightning Source UK Ltd.
Milton Keynes UK
17 March 2011

169453UK00007B/59/P